Diseases of Grape: Their Forecasting and Control

Diseases of Grape:
Their Forecasting and Control

Nasreen Akhter

RANDOM

RANDOM PUBLICATIONS
NEW DELHI (INDIA)

Diseases of Grape: Their Forecasting and Control

ISBN 978-93-5111-777-3

Published in 2016 in India by

RANDOM PUBLICATIONS

4376-A/4B, Gali Murari Lal, Ansari Road
New Delhi-110 002
Phone : +9111-43580356, 011-23289044, 011-43142548
e-mail: sales@randompublications.com,
info@randompublications.com, randomexports@gmail.com

Reprinted 2026

Type Setting by : Friends Media, Delhi-110089

Preface

Grape is an important fruit crop of the world contributing to economies of several countries due to its various uses including wine production. The production, productivity and economic gain substantially suffers due to vagaries of various disease attacking the grape crop world over. The present book on diseases of grape, their forecasting and control is an attempt to bring all the available information till date on this important topic in a global prospect and perspective.

Grapes can be grown in the backyard, but trellises will be needed to support them. American varieties are hardier and more suited for table use than European types. Suggested varieties include 'Niagara,' 'Steuben,' 'Swensen Red,' 'Fredonia,' 'Delaware,' 'Concord,' and 'Reliance' (seedless). Grapes require severe annual pruning early each spring to remain productive. Grapes flower and produce fruit only on one-year-old canes.

Rapid, accurate identification of diseases in the vineyard is key to preventing serious outbreaks and losses in yield and quality. However, the presence of a pathogen or disease does not necessarily mean that a treatment is required. The severity of diseases varies from year to year, depending primarily on weather conditions, the presence of inoculum (history of the disease) and the susceptibility of the vines. This means that a disease can be devastating one year and insignificant the next. The measures to be taken to prevent losses may therefore vary from season to season.

The purpose of this book is to aid in the identification of grape diseases and in pest management decision-making. Accurate identification of pests is critical to an effective management program that provides optimal control while minimizing pesticide use.

The book will prove its usefulness world over to the students of plant pathology, to research workers, extension workers and most importantly to grapevine growers who wishes to save their crop from vagaries of grape diseases.

– ***Author***

Contents

1

Introduction to Grape

GRAPE

A grape is a fruiting berry of the deciduous woody vines of the botanical genus *Vitis*. Grapes can be eaten raw or they can be used for making wine, jam, juice, jelly, grape seed extract, raisins, vinegar, and grape seed oil. Grapes are a non-climacteric type of fruit, generally occurring in clusters.

Fig. Grapes

Fig. "White" table grapes

HISTORY

The cultivation of the domesticated grape began 6,000–8,000 years ago in the Near East. Yeast, one of the earliest domesticatedmicroorganisms, occurs naturally on the skins of grapes, leading to the innovation of alcoholic drinks such as wine. The earliest archeological evidence for a dominant position of wine-making in human culture dates from 8,000 years ago in Georgia. The oldest winery was found in Armenia, dating to around 4000 BC. By the 9th century AD the city of Shiraz was known to produce some of the finest wines in the Middle East. Thus it has been proposed that Syrah red wine is named after Shiraz, a city in Persia where the grape was used to make Shirazi wine. Ancient Egyptian hieroglyphics record the cultivation of purple grapes, and history attests to the ancient Greeks, Phoenicians, and Romans growing purple grapes for both eating and wine production. The growing of grapes would later spread to other regions in Europe, as well as North Africa, and eventually in North America. In North America, native grapes belonging to various species of the *Vitis* genus proliferate in the wild across the continent, and were a part of the diet of many Native Americans, but were considered by European colonists to be unsuitable for wine. *Vitis vinifera*cultivars were imported for that purpose.

DESCRIPTION

Grapes are a type of fruit that grow in clusters of 15 to 300, and can be crimson, black, dark blue, yellow, green, orange, and pink. “White” grapes are actually green in color, and are evolutionarily derived from the purple grape. Mutations in two regulatory genes of white grapes turn off production of anthocyanins, which are responsible for the color of purple grapes. Anthocyanins and otherpigment chemicals of the larger family of polyphenols in purple grapes are responsible for the varying shades of purple in red wines. Grapes are typically an ellipsoid shape resembling a prolate spheroid.

GRAPEVINES

Fig. Concord is a variety of North American *labrusca*grape

Most grapes come from cultivars of *Vitis vinifera*, the European grapevine native to the Mediterranean and Central Asia. Minor amounts of fruit and wine come from American and Asian species such as:

- *Vitis labrusca*, the North American table and grape juice grapevines (including the Concord cultivar), sometimes used for wine, are native to the Eastern United States and Canada.
- *Vitis riparia*, a wild vine of North America, is sometimes used for winemaking and for jam. It is native to the entire Eastern U.S. and north toQuebec.
- *Vitis rotundifolia*, the muscadines, used for jams and wine, are native to the Southeastern United States from Delaware to the Gulf of Mexico.
- *Vitis amurensis* is the most important Asian species.

DISTRIBUTION AND PRODUCTION

According to the Food and Agriculture Organization (FAO), 75,866 square kilometers of the world are dedicated to grapes. Approximately 71% of world grape production is used for wine, 27% as fresh fruit, and 2% as dried fruit. A portion of grape production goes to producing grape juice to be reconstituted for fruits canned "with no added sugar" and "100% natural". The area dedicated to vineyards is increasing by about 2% per year.

There are no reliable statistics that break down grape production by variety. It is believed that the most widely planted variety is Sultana, also known as Thompson Seedless, with at least 3,600 km^2 (880,000 acres) dedicated to it. The second most common variety is Airén. Other popular varieties include Cabernet Sauvignon, Sauvignon blanc,Cabernet Franc, Merlot, Grenache, Tempranillo, Riesling, and Chardonnay.

TABLE AND WINE GRAPES

Fig. Wine grapes on the vine

Commercially cultivated grapes can usually be classified as either table or wine grapes, based on their intended method of consumption: eaten raw (table grapes) or used to make wine (wine grapes). While almost all of them belong to the same species, *Vitis vinifera*, table and wine grapes have significant differences, brought about through selective breeding. Table grape cultivars tend to have large, seedless fruit with relatively thin skin. Wine grapes are smaller, usually seeded, and have relatively thick skins (a desirable characteristic in winemaking, since much of the aroma in wine comes from the skin). Wine grapes also tend to be very sweet: they are harvested at the time when their juice is approximately 24% sugar by weight. By comparison, commercially produced "100% grape juice", made from table grapes, is usually around 15% sugar by weight.

SEEDLESS GRAPES

Seedless cultivars now make up the overwhelming majority of table grape plantings. Because grapevines are vegetatively propagated by cuttings, the lack of seeds does not present a problem for reproduction. It is an issue for breeders, who must either use a seeded variety as the female parent or rescue embryos early in development using tissue culture techniques. There are several sources of the seedlessness trait, and essentially all commercial cultivators get it from one of three sources: Thompson Seedless, Russian Seedless, and Black Monukka, all being cultivars of *Vitis vinifera*. There are currently more than a dozen varieties of seedless grapes. Several, such as Einset Seedless, Benjamin Gunnels's Prime seedless grapes, Reliance, and Venus, have been specifically cultivated for hardiness and quality in the relatively cold climates of northeastern United States and southern Ontario. An offset to the improved eating quality of seedlessness is the loss of potential health benefits provided by the enriched phytochemical content of grape seeds.

RAISINS, CURRANTS AND SULTANAS

Fig. Raisins

In most of Europe and North America, dried grapes are referred to as "raisins" or the local equivalent. In the UK, three different varieties are recognized, forcing the EU to use the term "dried vine fruit" in official documents. A *raisin* is any dried grape. While *raisin* is a French loanword, the word in French refers to the fresh fruit; *grappe* (from which the English*grape* is derived) refers to the bunch (as in *une grappe de raisins*). A *currant* is a dried Zante Black Corinth grape, the name being a corruption of the French *raisin de Corinthe* (Corinth grape). *Currant* has also come to refer to the blackcurrant and redcurrant, two berries unrelated to grapes. A *sultana* was originally a raisin made from Sultana grapes of Turkish origin (known as Thompson Seedless in the United States), but the word is now applied to raisins made from either white grapes or red grapes that are bleached to resemble the traditional sultana.

JUICE

Fig. Grape juice

Grape juice is obtained from crushing and blending grapes into a liquid. The juice is often sold in stores or fermented and made into wine, brandy, orvinegar. In the wine industry, grape juice that contains 7–23% of pulp, skins, stems and seeds is often referred to as "must". In North America, the most common grape juice is purple and made from Concord grapes, while white grape juice is commonly made from Niagara grapes, both of which are varieties of native American grapes, a different species from European wine grapes. In California, Sultana (known there as Thompson Seedless) grapes are sometimes diverted from the raisin or table market to produce white juice.

HEALTH CLAIMS

French paradox

Comparing diets among Western countries, researchers have discovered that although the French tend to eat higher levels of animal fat, the incidence of heart disease remains low in France. This phenomenon has been termed the French paradox, and is thought to occur from protective benefits of regularly consuming red wine. Apart from potential benefits of alcohol itself, including reduced platelet aggregation and vasodilation, polyphenols (e.g., resveratrol) mainly in the grape skin provide other suspected health benefits, such as:

- Alteration of molecular mechanisms in blood vessels, reducing susceptibility to vascular damage
- Decreased activity of angiotensin, a systemic hormone causing blood vessel constriction that would elevate blood pressure
- Increased production of the vasodilator hormone, nitric oxide (endothelium-derived relaxing factor)

Fig. Using grape leaves in cuisine (Dolma)

Although adoption of wine consumption is not recommended by some health authorities, a significant volume of research indicates moderate consumption, such as one glass of red wine a day for women and two for men, may confer health benefits. Emerging evidence is that wine polyphenols such as resveratrol provide physiological benefit, whereas alcohol itself may have protective effects on the cardiovascular system. More may be seen in the article Long-term effects of alcohol.

Resveratrol

Resveratrol is found in widely varying amounts among grape varieties, primarily in their skins and seeds, which, in muscadine grapes, have about one hundred times higher concentration than pulp. Fresh grape skin contains about 50 to 100 micrograms of resveratrol per gram.

Anthocyanins and other phenolics

Anthocyanins tend to be the main polyphenolics in purple grapes whereas flavan-3-ols (i.e. catechins) are the more abundant phenolic in white varieties.

Total phenolic content, a laboratory index of antioxidant strength, is higher in purple varieties due almost entirely to anthocyanin density in purple grape skin compared to absence of anthocyanins in white grape skin. It is these anthocyanins that are attracting the efforts of scientists to define their properties for human health. Phenolic content of grape skin varies with cultivar, soil composition, climate, geographic origin, and cultivation practices or exposure to diseases, such as fungal infections.

Red wine may offer health benefits more so than white because potentially beneficial compounds are present in grape skin, and only red wine is fermented with skins. The amount of fermentation time a wine spends in contact with grape skins is an important determinant of its resveratrol content. Ordinary non-muscadine red wine contains between 0.2 and 5.8 mg/L,depending on the grape variety, because it is fermented with the skins, allowing the wine to absorb the resveratrol. By contrast, a white wine contains lower phenolic contents because it is fermented after removal of skins. Wines produced from muscadine grapes may contain more than 40 mg/L, an exceptional phenolic content. In muscadine skins, ellagic acid, myricetin, quercetin, kaempferol, and trans-resveratrol are major phenolics. Contrary to previous results, ellagic acid and not resveratrol is the major phenolic in muscadine grapes. The flavonols syringetin, syringetin 3-O-galactoside, laricitrin and laricitrin 3-O-galactoside are also found in purple grape but absent in white grape.

Seed constituents

Biochemical and preliminary clinical studies have demonstrated potential biological properties of grape seed oligomeric procyanidins. For example, laboratory tests indicated a potential anticancer effect from grape seed extract. According to the American Cancer Society, "there is very little reliable scientific evidence available at this time that drinking red wine, eating grapes, or following the grape diet can prevent or treat cancer in people".

Grape seed oil from crushed seeds is used in cosmeceuticals and skincare products for perceived health benefits. Grape seed oil contains tocopherols (vitamin E) and high contents of phytosterols and polyunsaturated fatty acids such as linoleic acid, oleic acid, and alpha-linolenic acid.

Grape and raisin toxicity in dogs

The consumption of grapes and raisins presents a potential health threat to dogs. Their toxicity to dogs can cause the animal to develop acute renal failure (the sudden development of kidney failure) with anuria (a lack of urine production) and may be fatal.

Grape therapy

Grape therapy, also known as ampelotherapy (from Ancient Greek ìðåëïò *(ampelos)*, meaning "vine"), is a form of naturopathic medicine or alternative medicine that involves heavy consumption of grapes, including seeds, and parts of the vine, including leaves. Although there is some limited evidence of positive benefits from the consumption of grapes for health purposes, extreme claims, such as its ability to cure cancer, have been widely derided as "quackery".

RELIGIOUS SIGNIFICANCE

In the Bible, grapes are first mentioned when Noah grows them on his farm (Genesis 9:20–21). Instructions concerning wine are given in the book of Proverbs and in the book of Isaiah, such as in Proverbs 20:1 and Isaiah 5:20–25. Deuteronomy 18:3–5,14:22–27,16:13–15 tell of the use of wine during Jewish feasts. Grapes were also significant to both the Greeks and Romans, and their god ofagriculture, Dionysus, was linked to grapes and wine, being frequently portrayed with grape leaves on his head. Grapes are especially significant for Christians, who since the Early Church have used wine in their celebration of the Eucharist. Views on the significance of the wine vary between denominations. In Christian art, grapes often represent the blood of Christ, such as the grape leaves inCaravaggio's *John the Baptist*.

USE IN RELIGION

Grape juice, because of its non-alcoholic content, is commonly used by those Christians who oppose the partaking of alcoholic beverages, as the "cup" or "wine" in the Lord's Supper. The Catholic Church uses wine in the celebration of the Eucharist because it is part of the tradition passed down through the ages starting with Jesus Christ at the Last Supper, where Catholics believe the consecrated bread and wine *literally* become the body and blood of Jesus Christ, a dogma known as transubstantiation. Wine is used (not grape juice) both due to its strong Scriptural roots, and also to follow the tradition set by the early Christian Church.

The Code of Canon Law of the Catholic Church (1983), Canon 924 says that the wine used must be natural, made from grapes of the vine, and not corrupt. In some circumstances, a priest may obtain special permission to use grape juice for the consecration, however this is extremely rare and typically requires sufficient impetus to warrant such a dispensation, such as personal health of the priest. Although alcohol is permitted in Judaism, grape juice is sometimes used as an alternative for kiddush on Shabbat and Jewish holidays, and it has the same blessing as wine. Many authorities maintain that grape juice must be capable of turning into wine naturally in order to be used for kiddush. Common practice, however, is to use any kosher grape juice for kiddush.

VITIS

Vitis (grapevines) is a genus of about 60 species of vining plants in the flowering plant family Vitaceae. The genus is made up of species predominantly from the Northern hemisphere. It is economically important as the source of grapes, both for direct consumption of the fruit and for fermentation to produce wine. The study and cultivation of grapevines is called viticulture.

All vines share the same basic, physiological features. The roots anchor the vine to the soil and serve as the conduit where nutrients and water from the soil are absorbed. Along with the trunk or "permanent wood" features, the roots also serve as storage reserve ofcarbohydrates which the vine can use for energy in the winter after the leaves have fallen and are no longer conducting photosynthesis.

The function of photosynthesis in the grapevine is to produce glucose which can be combined with other molecules to form larger carbohydrates (such as cellulose) that can be used to create other structures in the vine, energy reserves for the plant and, for fruiting grapevines, can be concentrated in grape berries which contain the reproductive seeds of the vine and are more attractive to birds and other animals.

Other parts of the grapevine include the tendrils which the vine uses to support itself by clinging to surrounding structures (such as thetrellising of a vine-training system) and the shoots which sprout from the permanent wood and contain nodes where new leaves, flowers and tendrils can form. At the joint where leaves connect to the shoot are buds or "embryonic shoots" that contain the structures that will grow into the following years shoots, tendrils, flowers and leaves.

Most *Vitis* varieties are wind-pollinated with hermaphroditic flowers containing both male and female reproductive structures. These flowers are grouped in bunches called inflorescences. In many varieties, such as *Vitis vinifera*, each successfully pollinated flower becomes a grape berry with the inflorescence turning into a cluster of grapes. While the flowers of the grapevines are usually very small, the berries are often big, brightly colored with sweet flavors that attract birds and animals to disperse the seeds contained within the berries.

Grapevines usually only produce fruit on shoots that came from buds that were developed during the previous growing season. In viticulture, this is one of the principles behind pruning the previous years growth (or "One year old wood") that includes shoots that have turned hard and woody during the winter (after harvest in commercial viticulture). These vines will be pruned either into a cane which will support 8 to 15 buds or to a smaller spur which holds 2 to 3 buds.

Biology

Vitis is distinguished from other genera of Vitaceae by having petals which remain joined at the tip and detach from the base to fall together as a calyptra or 'cap'. The flowers are unisexual or modified to act functionally as unisexual, they are pentamerous with a hypogynous disk. The calyx is greatly reduced or nonexistent in most species and the petals are joined at the summit into one unit but separated at the base. Flower buds are formed later in the growing season and overwinter for blooming in spring of the next year. There are two types of flowers produced, sterile flowers with five long filaments and erect stamens with undeveloped pistils and fertile flowers with well-developed pistils and that have five undeveloped reflexed stamens. The fruit is a berry, normally produced with four or less per flower by way of aborted embryos, ovoid in shape and juicy.

Fig. Developing inflorescences of *Vitis vinifera*

In the wild, all species of *Vitis* are normally dioecious, but under domestication, variants with perfect flowers appear to have been selected.

Most *Vitis* species have 38 chromosomes (n=19), but 40 (n=20) in subgenus *Muscadinia*. In that respect the *Muscadinia* are the same as other *Vitaceae* such as *Ampelocissus*, *Parthenocissus*, and *Ampelopsis*.

USES

The fruit of several *Vitis* species are grown commercially for consumption as fresh grapes and for fermentation into wine. *Vitis vinifera* is the most important such species.

The leaves of several species of grapevine are edible and are used in the production of dolmades and Vietnamese lot leaves.

SPECIES

Most *Vitis* species are found in the temperate regions of the Northern Hemisphere in North America and Asia with a few in the tropics. The wine grape *Vitis vinifera* originated in southernEurope and southwestern Asia. The species occur in widely different geographical areas and show a great diversity of form.

Fig. *Vitis coignetiae* with autumn leaves

Their growth makes leaf collection challenging and polymorphic leaves make identification of species difficult. Mature grapevines can grow up to 48 cm in diameter at breast height and reach the upper canopy of trees more than 35 m in height (Everhart 2010, PDF)

Many species are sufficiently closely related to allow easy interbreeding and the resultant interspecific hybrids are invariably fertile and vigorous. Thus the concept of a species is less well defined and more likely represents the identification of different ecotypes of *Vitis* that have evolved in distinct geographical and environmental circumstances.

The exact number of species is not certain, with species in Asia in particular being poorly defined. Estimates range from 40 to more than 60. Some of the more notable include:

- *Vitis vinifera*, the European grapevine. Native to the Mediterranean and Central Asia.
- *Vitis labrusca*, the Fox grapevine, sometimes used for wine. Native to the Eastern United States and Canada.
- *Vitis riparia*, the Riverbank Grapevine, sometimes used for winemaking and for jam. Native to the entire Eastern U.S. and north toQuebec.
- *Vitis aestivalis*, the Summer Grape, native to the Eastern United States, especially the Southeastern United States.
- *Vitis rotundifolia* (syn. *Muscadinia rotundifolia*), the Muscadine, used for jams and wine. Native to the Southeastern United States fromDelaware to the Gulf of Mexico.
- *Vitis rupestris*, the Rock grapevine, used for breeding of Phylloxera resistant rootstock. Native to the Southern United States.

- *Vitis coignetiae* Crimson Glory Vine, a species from East Asia grown as an ornamental plant for its crimson autumn foliage.
- *Vitis amurensis*, native to the Asian continent, including parts of Siberia and China.
- *Vitis vulpina*, the Frost Grape, native to the Eastern United States, from Massachusetts to Florida, and west to Nebraska, Kansas, and Texas. Treated by some as a synonym of *V. riparia*.

There are many cultivars of grapevines; most are cultivars of *V. vinifera*.

Hybrid grapes also exist, and these are primarily crosses between *V. vinifera* and one or more of *V. labrusca*, *V. riparia* or *V. aestivalis*. Hybrids tend to be less susceptible to frost and disease (notably phylloxera), but wine from some hybrids may have a little of the characteristic "foxy" taste of *V. labrusca*.

The Latin word *Vitis* has feminine grammatical gender, and therefore species names with adjectival specific epithets take feminine forms, such as *V. vinifera*.

COMMERCIAL DISTRIBUTION

Fig. Vitis for producing Sherry at Jerez

According to the "Food and Agriculture Organization" (FAO), 75,866 square kilometres of the world is dedicated to grapes. Approximately 71% of world grape production is used for wine, 27% as fresh fruit, and 2% as dried fruit. A portion of grape production goes to producing grape juice to be used as a sweetener for fruits canned "with no added sugar" and "100% natural". The area dedicated to vineyards is increasing by about 2% per year.

The following list of top wine-producers shows the corresponding areas dedicated to grapes for wine making:

- Spain 11,750 km^2
- France 8,640 km^2

- Italy 8,270 km^2
- Turkey 8,120 km^2
- United States 4,150 km^2
- Iran 2,860 km^2 (No wine production. Grapes are used as a table-fruit. The leaves are widely used in cooking, as wrappers.)
- Romania 2,480 km^2
- Portugal 2,160 km^2
- Argentina 2,080 km^2
- Australia 1,642 km^2
- Lebanon 1,221 km^2

DOMESTIC CULTIVATION

Grapevines are widely cultivated by gardeners, and numerous suppliers cater specifically for this trade. The plants are valued for their decorative foliage, often colouring brightly in autumn; their ability to clothe walls, pergolas and arches, thus providing shade; and their fruits, which may be eaten as dessert or provide the basis for homemade wines. Popular varieties include:-

'Schiava Grossa' (black dessert)	'Foster's Seedling' (white dessert)	'Phoenix' (white wine)
'Muscat of Alexandria' (white dessert)	'Müller-Thurgau' (white wine)	'Chardonnay' (white wine)
Buckland Sweetwater' (white dessert)	'Seyval blanc' (white wine)	'Pinot noir' (black wine)
		'Regent' (black wine)

The following varieties, grown primarily for their ornamental qualities, have gained the Royal Horticultural Society's Award of Garden Merit:-

'Brant' (black dessert)	*Vitis coignetiae* (ornamental)
'New York Muscat' (black dessert)	'Purpurea' (ornamental)

PESTS AND DISEASES

Fig. 'Palatina', a German grape

Phylloxera is an American root aphid that devastated *V. vinifera* vineyards in Europe when accidentally introduced in the late 19th century. Attempts were made to breed in resistance from American species, but many winemakers didn't like the unusual flavour profile of the hybrid vines. Fortunately, *V. vinifera* grafts readily onto rootstocks of the American species, and most commercial production of grapes now relies on such grafts.

The Black vine weevil is another root pest.

Grapevines are used as food plants by the larvae of some Lepidoptera species.

SYMBOLISM

The grapevine (typically *Vitis vinifera*) has been used as a symbol since ancient times. In Greek mythology, Dionysus (called Bacchus by the Romans) was god of the vintage and, therefore, a grapevine with bunches of the fruit are among his attributes. His attendants at the Bacchanalian festivals hence had the vine as an attribute, together with thethyrsus, the latter often entwined with vine branches. For the same reason, the Greek wine cup (cantharos) is commonly decorated with the vine and grapes, wine being drunk as a libation to the god. In Christian iconography, the vine also frequently appears. It is mentioned several times in the New Testament. We have the parable of the kingdom of heaven likened to the father starting to engage laborers for his vineyard. The vine is used as symbol of Jesus Christ based on his own statement, "I am the vine." In that sense, a vine is placed as sole symbol on the tomb of Constantia, the sister of Constantine the Great, and elsewhere. In Byzantine art, the vine and grapes figure in early mosaics, and on the throne ofMaximianus of Ravenna it is used as a decoration.

The vine as symbol of the chosen people is employed several times in the Old Testament. The vine and wheat ear have been frequently used as symbol of the blood and flesh of Christ, hence figuring as symbols (bread and wine) of the Eucharist and are found depicted on ostensories. Often the symbolic vine laden with grapes is found in ecclesiastical decorations with animals biting at the grapes. At times, the vine is used as symbol of temporal blessing.

RAISIN

Fig. The natural sugars in grapes crystallize during drying

Fig. Raisins (Sultanas)

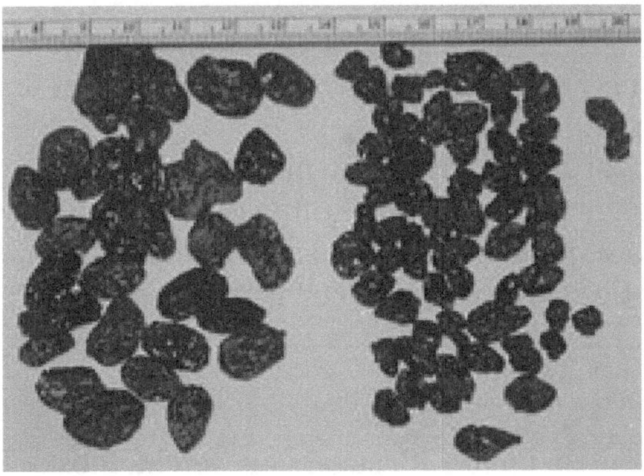

Fig. California seedless grape raisins on the left and California Zante currants on the right, along with a metric ruler for scale.

A raisin is a dried grape. Raisins are produced in many regions of the world and may be eaten raw or used in cooking, baking andbrewing. In the United Kingdom, Ireland, New Zealand and Australia the word "raisin" is reserved for the dark-coloured dried large grape, with "sultana" being a golden-coloured dried grape, and "currant" being a dried small Black Corinth grape.

ETYMOLOGY

The word *raisin* dates back to Middle English and is a loanword from Old French; in modern French, *raisin* means "grape", while a dried grape is a *raisin sec*, or "dry grape". The Old French word in turn developed from the Latin word *racemus*, "a bunch of grapes".

VARIETIES

Raisin varieties depend on the type of grape used, and are made in a variety of sizes and colors including green, black, blue, purple and yellow. Seedless varieties include the sultana (also known as Thompson Seedless in the USA) and Flame grapes. Raisins are traditionally sun-dried, but may also be water-dipped and artificially dehydrated.

Fig. A variety of raisins from different grapes

"Golden raisins" are treated with sulfur dioxide after drying to give them their golden color.

Black Corinth or Zante currant are miniature raisins that are much darker and have a tart, tangy flavor. They are often called currants.Muscat raisins are large compared to other varieties, and also sweeter.

Several varieties of raisins produced in Asia are available in the West only at ethnic grocers. Monukka grapes are used for some of these.

NUTRITION

Raisins can contain up to 72% sugars by weight, most of which is fructose and glucose. They also contain about 3% protein and 3.7%–6.8% dietary fiber. Raisins, like prunes and apricots, are also high in certain antioxidants, but have a lower vitamin Ccontent than fresh grapes. Raisins are low in sodium and contain no cholesterol.

Data presented at the American College of Cardiology's 61st Annual Scientific Session in 2012 suggests that, among individuals with mild increases in blood pressure, the routine consumption of raisins (three times a day) may significantly lower blood pressure, especially when compared to eating other common snacks.

TOXICITY IN PETS

Raisins can cause renal failure in dogs. The cause of this is not known.

SUGARS

Raisins are sweet due to their high concentration of sugars (about 30% fructose and 28% glucose by weight). The sugars cancrystallise inside the fruit when stored after a long period, making the dry raisins gritty, but that does not affect their usability. These sugar grains can be dissolved by blanching the fruit in hot water or other liquids.

Grades of raisins in the U.S.

Grade A: The color is good, and the flavor is characteristic of raisins. These raisins show development characteristics indicative of the fact that they are prepared from well-matured grapes (containing no less than 80% water by weight). The processed raisins contain less than 19% moisture by weight.

Grade B: The color and flavor is *reasonably* good. These raisins show development characteristics that highlight the fact that the raisins were prepared from *reasonably* well-matured grapes containing at least 70% water by weight. These raisins also contain less than 19% moisture, by weight.

Grade C: The color and flavor are *fairly* good. These raisins show development characteristics of raisins prepared from *fairly* well-matured grapes containing at least 55% water by weight. These raisins also contain less than 19% moisture, by weight.

Substandard: Raisins that fail to meet the standards of Grade C.

RAISIN PRODUCTION

Raisins are produced commercially by drying harvested grape berries. In order for a grape berry to dry, water inside the grape must be removed completely from the interior of the cells onto the surface of the grape where the water droplets can evaporate. However, this diffusion process is very difficult because the grape skin contains wax in its cuticle, which prevents the water from passing through. In addition to this, the physical and chemical mechanisms located on the outer layers of the grape are adapted to prevent water loss.

There are 3 steps to commercial raisin production including a pre-treatment, drying, and post-drying process.

Pre-treatment

Pre-treatment is a necessary step in raisin production in order to ensure the increased rate of water removal during the drying process. A faster water removal rate decreases the rate of browning and helps to produce more desirable raisins. The historical method of completing this process was developed in the Mediterranean and Asia Minor areas by using a dry emulsion cold dip made of potassium carbonate and ethyl esters of fatty acids. This dip was shown to increase the rate of water loss by two- to three-fold.Recently, new methods have been developed such as exposing the grapes to oil emulsions or dilute alkaline solutions. These methods can encourage water transfer to the outer surface of grapes which helps to increase the efficiency of the drying process.

Drying

There are three types of drying methods: sun drying, shade drying, and mechanical drying. Sun drying is an inexpensive process; however, many concerns such as environmental contamination, insect infections and microbial deterioration are inevitable and the resulting raisins are often of low quality. Additionally, sun drying is a very slow process and may not produce the most desirable raisins.

Mechanical drying can be done in a safer and more controlled environment where rapid drying is guaranteed. One type of mechanical drying is to use microwave heating. Water molecules in the grapes will absorb microwave energy resulting in rapid evaporation. Microwave heating often produces puffy raisins.

Post-drying processes

After the drying process is complete, raisins are sent to processing plants where they are cleaned with water to remove any foreign objects that may have become embedded during the drying process. Stems and off-grade raisins are also removed.

The washing process may cause rehydration so another drying step is completed after washing to ensure that all moisture has been removed.

All steps in the production of raisins are very important in determining the quality of raisins. Sometimes sulfur dioxide is applied to raisins after the pre-treatment step and before drying in order to decrease the rate of browning caused by the reaction between polyphenol oxidase and phenolic compounds. Sulfur dioxide also helps to preserve flavour and prevent the loss of certain vitamins during the drying process.

Nutrition and health

Raisins can be eaten as a nutritious snack, rich in dietary fiber, carbohydrates with a low glycemic index, minerals, vitamins and other micronutrients.

Their fat and cholesterol content is low. The overall nutritional value of raisins means that they are recommended as a snack for both weight control and for maintaining good human health because they help the control of glucose and cholesterol, the good functioning of the digestive system and the regulation of blood pressure.

Replacing unhealthy snacks by raisins in usual and moderate quantity can improve health biomarkers in patients with controlled Type 2 Diabetes. It has been found that adoption of this dietary habit may reduce diastolic blood pressure and increase the levels of plasma antioxidants in type 2 diabetic patients. Corinthian raisins are a moderate glycemic index fruit. They can be consumed in small amounts even by diabetic patients instead of sweets in a balanced diet.

Antioxidants in Greek raisins may reduce the risk for malignancies in the stomach and colon. Thus, in a balanced diet, their antioxidants can help maintain the health of the digestive system.

GRAPE THERAPY

Grape therapy, also known as ampelotherapy (from Ancient Greek ìðåëïò *(ampelos)*, meaning "vine"), is a form of naturopathic alternative medicine that involves heavy consumption of grapes, including seeds, and parts of the vine, including leaves. Although there is some limited evidence of positive benefits from the use of grape therapy for health purposes, extreme claims, such as its ability to cure cancer, have been widely derided as "quackery".

BACKGROUND

Grapes have been associated with health for many centuries. They contain organic acids (tartaric, malic, citric, tannic), B group vitamins, vitamins P, PP, and C, calcium,carotene, magnesium salts, phosphorus and potassium, as well as trace elements such as copper, iron, manganese, sulphur and zinc. Proanthocyanidins, first extracted from grape seeds in 1970, and resveratrol, found in grape skins, are currently being studied for possible uses in the prevention and treatment of cancer and other illnesses, and for anti-ageing. Proanthocyanidins are powerful antioxidants that can block the action of free radicals.

Advocates of grape therapy argue that these antioxidants inhibit the development of some types of cancer, and can be used in treating a variety of other medical conditions such as arthritis and diabetes. Grapes are said to be beneficial for protecting against cardiovascular disease and, because of their potassium and glucose content, they are argued to be useful for treatment of heart disease by strengthening muscle tone, improving heart function and having a positive impact on convalescence. Alternative medicine practitioners have at times also recommended grapes and parts of the vine for high blood pressure, menopause, varicose veins, high cholesterol, skin rashes, and urination problems.

The grape diet was promoted in the past as a treatment to flush toxins from the body. The grape cure is also prescribed for a decline in nutrition, especially with patients with early stage of tuberculosis, for diseases of the liver, kidneysand lungs and for gout and secondary anaemia and, in addition, for mental disease, nervous disorders, and sexual problems. Recent research by scientists at UCLA has found that combining resveratrol with benzoyl peroxide may improve that drug's ability to kill the bacteria associated with acne. The presence of resveratrol on the skin of red grapes has prompted some doctors to recommend that adults drink reasonable quantities of red wine to combat heart disease. Resveratrol in red wine is suggested as an explanation for the "French paradox," i.e. the low occurrence of heart disease among the French, even though they eat a high-fat diet.

HISTORY

The documentation of grape cures was first seen in a publication by V.N. Dmitriev called "Treatment With Grapes in Yalta on the Southern Crimean Shore" (1878). In the USSR, the principles of the grape cure were developed in the 1920s by a group of physicians of the Semashko Institute (Yalta), headed by A.V. D'iakov.

Ampelotherapy continues to be offered in health clinics and spas, particularly in Europe, together with Vinotherapy, a cosmetic treatment that involves rubbing grapes into the skin. Johanna Brandt, a South African dietician, popularized the grape diet as a treatment for cancer from 1925. She published about twenty pamphlets on the subject of natural remedies for health problems with her best-known publication being *The Grape Cure*, which is said to have been written after Brandt claimed to have cured herself of stomach cancer by following the diet. The book was republished in 1989 as *How to Conquer Cancer, Naturally*, including an endorsement of Brandt's work by Benedict Lust, who is commonly referred to as "the father of naturopathy." The book may have been inspired by Arnold Ehret, a contemporary, who taught a Grape Cure course.

TREATMENTS

Fig. An abandoned Ampelotherapy clinic on the outskirts of Rome, Italy

The leaves, stems, and grape itself, including seeds and skins, are used in herbal remedies. Treatments offered have varied from a diet that consists of eating large quantities of grapes together with other food to those that involve consuming nothing but grapes. One recommended course of ampelotherapy has to last 1–2 months.

It is recommended to eat 2 kilos of grapes with their seed three times per week. This intake should be accompanied by a diet that limits the intake of smoked meat, milk and all dairy products and alcohol. The diet proposed by Johanna Brandt recommended fasting for two or three days, consuming only cold water, followed by a diet of only grapes and water for one to two weeks, with seven meals a day. Fresh fruits, tomatoes, and sour milk or cottage cheese are then introduced to the diet followed by raw vegetables.

CRITICISMS

Available scientific evidence does not support claims that a diet of grapes is alone effective for treating cancer or any other disease. Some laboratory evidence suggests that certain chemicals in grapes, their seeds and skins may help prevent heart disease and cancer, but more research is needed on humans to understand the possible long-term benefits. The Brandt diet, in particular, has been described as "quackery" by Barrett who notes that The American Cancer Society reviewed the *The Grape Cure* in 1965, 1971, 1974, and 2000 and found no evidence of benefit against human cancer or any other disease. Even worse, he argues that the diet can cause constipation, diarrhea, cramps, and weight loss that is undesirable for cancer patients An exclusive grape diet is unhealthy and does not supply the body with adequate amounts of protein and important nutrients.

GRAPE SEED EXTRACT

Grape seed extracts are industrial derivatives from whole grape seeds that have a great concentration of vitamin E, flavonoids, linoleic acid and phenolic procyanidins (also known as OPC or oligomeric procyanidins). The typical commercial opportunity of extracting grape seed constituents has been for chemicals known as polyphenols havingantioxidant activity in vitro.

POSSIBLE HEALTH BENEFITS

According to the American Cancer Society, "there is very little reliable scientific evidence available at this time that drinking red wine, eating grapes, or following the grape diet can prevent or treat cancer in people".

A polyphenol contained in grape seeds is resveratrol, which is under study for its possible effect on cancer cell growth, proliferation or apoptosis, among other potentialchemopreventive mechanisms.

Other preliminary research on disease models

- skin and wounds – OPCs induced vascular endothelial growth factor and accelerated healing of injured skin in mice
- teeth – seed phenolics may inhibit oral sugar metabolism and retard growth of certain bacteria that cause dental caries

- bones – grape seed extracts enhanced bone density and strength in experimental animals
- in vitro cancer studies – grape seed proanthocyanidins decreased tumor numbers and reduced the malignancy of papillomas
- ultraviolet damage – dietary proanthocyanidins are under study for mechanisms against carcinogenesis and sunscreen protection
- anti-viral effects
- antibacterial properties
- liver function
- blood flow and fluid balance

One clinical trial with adults having coronary disease or cardiac risk factors concluded that: "Four weeks of muscadine grape seed supplementation in subjects with increased cardiovascular risk did not produce a statistically significant increase in brachial flow-mediated vasodilation or a significant change in other biomarkers of inflammation, lipid peroxidation, or antioxidant capacity. However, the muscadine grape seed supplement did result in a significant increase in resting brachial diameter. The clinical significance of the effect on resting diameter is not yet established."

A meta-analysis of randomized controlled trials concluded that "grape seed extract appears to significantly lower systolic blood pressure and heart rate, with no effect on lipid orC-reactive protein levels." The US National Center for Complementary and Integrative Health reported that oral administration of grape seed extract was well tolerated in people over 8 weeks. In one completed clinical trial, grape seed extract did not alleviate the hardening of breast tissue in female patients undergoing radiation therapy to treat breast cancer.

DOSAGE, PRECAUTIONS AND INTERACTIONS

Oral grape seed extract is used in capsules or tablets usually containing 50 mg or 100 mg. Insufficient scientific information is known, however, about how long-term use of grape seed extract might affect health or any disease.

Side-effects and cautions, other NCCIH advisories

- In general, grape seed extract is well tolerated when taken by mouth, although it is better tolerated when encapsulated, as its taste is bitter. It has been used safely for up to 8 weeks in clinical trials
- Side-effects most often include headache, a dry, itchy scalp, dizziness, or nausea
- Interactions between grape seed extract and medicines or other supplements have not been carefully studied

Because of the possible action of proanthocyanidins on limiting platelet adhesion, grape seed extract may act as a blood-thinner, increasing clotting time.

AROMATASE INHIBITOR

Grape seed extract is also an aromatase inhibitor *in vitro*, i.e. it may suppress the conversion of testosterone to estradiol.

GRAPE SEED OIL

Grape seed oil (also called grapeseed oil or grape oil) is pressed from the seeds of grapes, and is thus an abundant by-productof winemaking.

USES

Cooking

Grape seed oil has a moderately high smoke point of approximately 216 °C (421 °F). Due to its clean, light taste, and highpolyunsaturated fat content, it may be used as an ingredient in salad dressings and mayonnaise and as a base for oil infusions ofgarlic, rosemary, or other herbs or spices. It is widely used in baked goods, pancakes, and waffles. It is sprayed on raisins to help them retain their flavor. The metabolic energy density of grape seed oil is typical of vegetable oils: approximately 3,700 kJ (880 kcal) per 100 g, or 500 kJ (120 kcal) per 15 ml tablespoon.

Cosmetics

Grape seed oil is a preferred cosmetic ingredient for controlling moisture of the skin. Light and thin, grape seed oil leaves a glossy film over skin when used as a carrier oil for essential oils in aromatherapy. It contains more linoleic acid than many other carrier oils.

Grape seed oil is also used as a lubricant for shaving and as a growth and strengthening treatment for hair.

Potential medicinal benefits

Grape seed oil may provide some health benefit. A 1993 study supports the claim that grape seed oil increases high-density lipoprotein (HDL-C or "good cholesterol") levels and reduces LDL levels.

Although grape seeds contain antioxidants and other biologically active compounds, the cold-pressed grape seed oil contains negligible amounts due to their insolubility in lipids. For instance, sufficiently high amounts of resveratrol occur in grape seed for it to be extracted commercially, yet it is almost entirely absent in the grape seed oil.

Potential medicinal complications

Grapeseed oil has sometimes been found to contain dangerous levels of polycyclic aromatic hydrocarbons because of direct contact with combustion gases during the drying process.

COMPOSITION

The following table lists a typical fatty acid composition of grape seed oil:

Acid	Type	Percentage
Linoleic acid	ω−6 unsaturated	69.6%
Oleic acid	ω−9 unsaturated	15.8%
Palmitic acid (Hexadecanoic acid)	Saturated	7%
Stearic acid (Octadecanoic acid)	Saturated	4%
Alpha-linolenic acid	ω−3 unsaturated	0.1%
Palmitoleic acid (9-Hexadecenoic acid)	ω−7 unsaturated	less than 1%

Grape seed oil also contains 0.8 to 1.5% unsaponifiables rich in phenols (tocopherols) and steroids (campesterol, beta-sitosterol,stigmasterol). Grapeseed oil contains small amounts of vitamin E, but safflower oil, cottonseed oil, or rice bran oil contain greater amounts. Grapeseed oil is high in polyunsaturates and low in saturated fat.

2

Grapes: Nutritional Benefits for Health

It may surprise you to hear that the humble grape is actually considered to be a berry. Seventy-two million tons of these berries are grown each year worldwide, with most of them being used to produce wine - 7.2 trillion gallons of wine per year, to be exact.

POSSIBLE HEALTH BENEFITS OF GRAPES

Consuming fruits and vegetables of all kinds has long been associated with a reduced risk of heart disease, diabetes, cancer and other conditions.

Many studies have shown that increasing consumption of plant foods like grapes decreases the risk of obesity and overall mortality. Grapes also have some special components that make them even more essential to our health, giving them "super food" status and reducing the risk of the following conditions:

CANCER

Grapes contain powerful antioxidants known as polyphenols, which may slow or prevent many types of cancer, including esophageal, lung, mouth, pharynx, endometrial, pancreatic, prostate and colon. The resveratrol found in red wine famous for heart health is a type of polyphenol found in the skins of red grapes.

HEART DISEASE

The flavonoid quercetin is a natural anti-inflammatory that appears to reduce the risk of (atherosclerosis and protect against the damage caused by low-density lipoprotein (LDL) cholesterol in animal studies. Quercetin may have the additional bonus of anti-cancer effects; however more studies are needed using human subjects before these results can be confirmed. The high polyphenol content in grapes may also reduce the risk of cardiovascular disease (CVD) by preventing platelet build-up and reducing blood pressure via anti-inflammatory mechanisms.

The fiber and potassium in grapes also support heart health. An increase in potassium intake along with a decrease in sodium intake is the most important dietary change that a person can make to reduce their risk of cardiovascular disease, according to Mark Houston, MD, MS, an associate clinical professor of medicine at Vanderbilt Medical School and director of the Hypertension Institute at St Thomas Hospital in Tennessee.

In one study, those who consumed 4069 mg of potassium per day had a 49% lower risk of death from ischemic heart disease compared with those who consumed less potassium (about 1000 mg per day). High potassium intakes are also associated with a reduced risk of stroke, protection against loss of muscle mass, preservation of bone mineral density and reduction in the formation of kidney stones.

HIGH BLOOD PRESSURE

As noted above, potassium has many benefits for the body. It may be that a low potassium intake is just as big of a risk factor in developing high blood pressure as a high sodium intake.4 Because of their high potassium content, grapes are recommended to those with high blood pressure to help negate the effects of sodium in the body.

According to the National Health and Nutrition Examination Survey, fewer than 2% of US adults meet the daily 4700 mg recommendation for potassium. Also of note, a high potassium intake is associated with a 20% decreased risk of dying from all causes.

Constipation

Eating foods that are high in water content like grapes, watermelon and cantaloupe can help to keep you hydrated and your bowel movements regular. Grapes also contain fiber, which is essential for minimizing constipation.

Allergies

Because of the anti-inflammatory effects of quercetin, consuming grapes may help to alleviate symptoms of allergies including runny nose, watery eyes and hives. There have been no human studies done to prove this theory.

Diabetic neuropathy and retinopathy

A few studies have shown promise that resveratrol can protect against diabetic neuropathy and retinopathy, conditions caused by poorly controlled diabetes where vision is severely affected. One study in which diabetic rats were treated with resveratrol for two weeks found that it reduced the effects of neural changes and damage associated with diabetic neuropathy. Researchers have also found resveratrol to be beneficial for treating Alzheimer's disease, relieving hot flashes and mood swings associated with menopause and improving blood glucose control, however large studies using human subjects are still needed to confirm these findings.

HEALTH BENEFITS

Fruit is an essential part of your daily diet supplying essential vitamins and minerals that contribute to overall good health. The latest USDA Dietary Guidelines recommend the following: fill half your plate with fruit and vegetables. Grapes' nutritional benefits make them an excellent choice for a more satisfying, healthier diet.

WHAT ARE THE HEALTH BENEFITS OF GRAPES?

Grapes are loaded with nutrients and contain antioxidants. Here are seven great reasons to snack on grapes or add them to your meals:

Heart Disease

The polyphenols found in grapes, have been shown to reduce the risk of heart disease and support a healthy heart by promoting relaxation of blood vessels to help maintain healthy blood flow and function.

Weight Loss

When you're striving for weight loss, the goal is to eat low-energy, dense foods like grapes. Meaning, you want to eat a greater volume of lower calorie foods. This helps you feel fuller on fewer calories.

Eye Health

Research suggests that regular grape consumption may play a role in eye health by protecting the retina from deterioration. Specifically a grape-enriched diet resulted in a protective effect on retinal structure and function.

High Blood Pressure

Grapes are naturally full of potassium, an essential dietary mineral and major electrolyte. Your average seedless green grapes contain right around 175 milligrams of potassium per cup, red grapes nearly 290 milligrams, and black grapes 200 milligrams. It may be that low potassium intake is just as big of a risk factor in developing high blood pressure as a high sodium intake. Plus, potassium can help combat excessive sodium in the body. Grapes are recommended to those with high blood pressure to help negate the effects of sodium in the body.

Health-Protecting Antioxidants

Grapes are rich in health-protecting antioxidants, including resveratrol and flavonoids. These antioxidants are found mainly in the skin, stem, leaf and seeds of grapes. Some research suggests that whole grapes deliver the same amount of antioxidants that are in grape juice and wine but have the added benefit of providing dietary fiber.

Blood Sugar and Diabetes

Grapes do not tend to highly elevate your blood sugar when eaten, nor do they highly increase insulin levels. Since high insulin levels cause insulin resistance, choosing foods with a lower insulin response rate, like grapes, will help to prevent you from developing the insulin resistance characteristic of Type 2 diabetes.

Colon Health

In a small study of colon cancer patients, those who consumed a grape-enriched diet were able to inhibit certain genes that promote tumor growth in the colon. This benefit was observed in the healthy tissue of the subjects' colons, not the cancerous, indicating a potential role for grapes in helping to maintain a healthy colon.

THE NUTRITIONAL FOOD VALUE OF GRAPES

Whether you're eating red or green grapes, the nutritional food value of grapes can help you improve your health in a variety of different ways. Though you *do* have the option of eating grapes either cooked or raw, you probably want to stick with eating raw grapes as they taste better *and* contain all of the nutrients that you'll need to stay healthy. Raw grapes are also a quick and easy snack to substitute into your diet over less healthy options like cookies and candies. They contain a small amount of natural sugar that will give you energy and a boost during the day without necessarily hurting your waistline.

In addition to being a convenient snack for you, grapes are also an excellent source of several different vitamins and nutrients that help keep you healthy. They can even prevent and help treat certain diseases. Here are just a few of the vitamins andnutrients presents in a single serving of grapes:

VITAMIN C

Grapes contain high levels of vitamin C. Vitamin C is a powerful antioxidant that assists your body in ridding itself of free radicals within the body. These radicals can attack your cells, weaken your immune system and allow your body to catch certain diseases. The common cold is typically a result of lowered levels of vitamin C in your body. Try eating grapes to supplement your body with vitamin C.

VITAMIN B1

If you're looking to get an energy boost, grapes help you by providing your body with vitamin B1. Vitamin B1 is a vitamin that will give your body energy. Aside from doing that, it will also help keep your nervous system to function properly.

FLAVANOIDS

Though you can get flavanoids in leafy, green vegetables, too, grapes are a great source of them. The grapes found in flavanoids reduce your body's chance of suffering from blood clots. If they become enough of a problem, blood clots can cause strokes and also put you at a higher risk for heart attacks and other forms of heart disease. The flavanoids in grapes reduce the risk of these clots, which promotes good blood circulation throughout your body.

POTASSIUM

If you suffer from high blood pressure, grapes could help to lower it by providing your body with potassium. Potassium also helps fight your body's chances of undergoing a stroke. If you suffer from a lack of energy, you could also be suffering from a potassium deficiency. Try a single serving of grapes everyday to assist you with this.

MANGANESE

In addition to the vitamins and nutrients listed above, grapes also have high levels of manganese in them. Manganese helps the body build stronger bones and also builds the connective tissue throughout your body. It also assists the body in calcium absorption, which is very important for those looking to build healthy bones.

The nutritional food value of grapes can help you and your body to grow and remain healthy. Try incorporating grapes into your diet to get more energy, fight heart disease and increase your chances of living longer.

FIVE HEALTH BENEFITS OF CONCORD GRAPES

Dark purple concord grapes are rich in antioxidants and can help keep your immune system healthy. Try them in this healthy Waldorf salad.

Concord grapes actually have the power to keep us youthful because they contain high concentrations of antioxidants that protect our skin, brain and heart from inflammation and aging. The healthiest part of the grape is the skin, which is packed with not one but 19 different types of health-supporting nutrients.

Here are five huge reasons to enjoy concord grapes:

1. Decrease blood pressure with concord grapes: Concord grapes contain many flavonoids, including resveratrol, which can help decrease blood pressure by improving the fluidity of the blood. Resveratrol also relaxes the arterial walls to allow adequate circulation and decrease the pressure in the arteries.
2. Concord grapes are anti-inflammatory: Concord grapes contain a variety of polyphenols that can reduce the inflammatory response in the body. These grapes also have the antioxidant ability to mop up free radicals in the body and keep you looking younger, longer.

3. Eat grapes to stay sharp: Many degenerative diseases, including Alzheimer's and dementia, begin with the accumulation of certain proteins and foreign bodies in the brain. Concord grapes have been shown to prevent the accumulation of these potentially harmful substances. This helps keep your brain sharp and your nervous system healthy.
4. Boost your immunity with concord grapes: Concord grapes have been shown to increase the production of the cells in our body that fight off disease and foreign invaders. Boosting these natural killer cells and T-cells will keep your immune system in tip-top shape during cold and flu season.
5. Concord grapes maintain healthy breast tissue: Concord grapes contain resveratrol, which protects the DNA in our cells against mutations that occur. Studies show that resveratrol is specifically beneficial in protecting breast tissue from the mutations that can take place in certain breast cancers.

Waldorf salad

Waldorf salad is often served with a creamy dressing, but in this incredibly healthy version, the sweetness of the grapes speaks for itself. Walnut oil is a source of omega-3 fatty acids that augment the heart and brain benefits of the grapes.

Ingredients:; Directions:

1. Toss together the apples, celery, grapes, and walnuts in a large bowl; set aside.
2. In a small bowl, whisk together the walnut oil, apple cider vinegar and salt.
3. Toss together and serve.

NUTRITIONAL BREAKDOWN OF GRAPES

The average serving size for grapes is about one cup, or 32 grapes. As a useful measure, you can use the size of your fist to estimate the proper portion size for grapes.

One cup of red or green grapes contains 104 calories, 1.09 grams of protein, 0.24 grams of fat, 1.4 grams of fiber, 4.8 milligrams ofvitamin C, 10 micrograms of vitamin A, 288 milligrams of potassium, 0.54 milligrams of iron and 3 micrograms of folate. Grapes are high in water content and good for hydration. High water-content fruits and vegetables are nutrient dense, meaning they provide a large amount of essential nutrients while containing few calories. Grapes contain 70 milliliters of fluid per cup.

Grapes are high in antioxidants important for eye health such as lutein and zeaxanthin, and red grapes contain the phytochemical resveratrol in their skins, the antioxidant synonymous with wine known to lend protection from several chronic diseases and conditions. Grapes also boast the power of the flavonoids myricetin and quercetin which help the body to counter-act harmful free radical formation.

INCORPORATING MORE GRAPES INTO YOUR DIET

The Environmental Working Group produces a list each year of fruits and vegetables with the highest levels of pesticide residue, known as the Dirty Dozen. Grapes are high on the list of produce that the EWG suggests that you buy in the organic version to ensure a lower risk of pesticide exposure.

Grapes are available year round. Select grapes that are tight to the touch and free of wrinkles. Grapes are best stored in the refrigerator and should be washed before eaten.

The best way to consume grapes is fresh right out of your hand! Most grape jellies/spreads, juices and wine have added sugars, which tack on additional calories.

Here are some handy tips to incorporate more grapes into your diet:

- Slice grapes in half and add them to your chicken salad.
- Make your own fruit cocktail with fresh fruit and include grapes, pineapple, sliced peaches and strawberries. Drizzle a small amount of honey on top of the fruit mixture for an extra sweet treat.
- Freeze grapes and eat them as a snack or dessert on a hot day.
- Have a quick 100-calorie snack with 1 cup of fresh grapes.

Risks and precautions associated with grapes

Beta-blockers, a type of medication most commonly prescribed for heart disease, can cause potassium levels to increase in the blood. High potassium foods such as bananas should be consumed in moderation when taking beta-blockers. Consuming too much potassium can be harmful for those whose kidneys are not fully functional. If your kidneys are unable to remove excess potassium from the blood, it could be fatal.

When drinking wine to reap it's health benefits drink no more than one drink per day for women or two drinks per day for men.

WHAT'S NEW AND BENEFICIAL ABOUT GRAPES

- Several grape phytonutrients are now believed to play a role in longevity. At the top of the list in this area of research is resveratrol (a stilbene phytonutrient present mostly in grape skins, but also in grape seeds and grape flesh). Resveratrol has recently been shown to increase expression of three genes all related to longevity. (These three genes are SirT1s, Fox0s, and PBEFs.) Interestingly, some researchers have shown a parallel between activation of these longevity genes by resveratrol and activation by calorie-restricted diets. In aging and longevity research, our ability to get optimal nutrition for the fewest possible amount of calories is related to our longevity, and the more we can decrease our calories while staying optimally nourished, the better our chances of healthy aging and longevity.
- Grapes have long been classified as a low glycemic index (GI) food, with GI values ranging between 43-53. But having a low GI value is not necessarily the same as having blood sugar benefits. In the case of grapes, recent studies have shown that the low GI value of grapes is also a good indicator of this fruit's blood sugar benefits. Better blood sugar balance, better insulin regulation, and increased insulin sensitivity have now been connected with intake of grape juices, grape extracts, and individual phytonutrients found in grapes.
- The 2014 edition of the Shopper's Guide to Pesticides by the Environmental Working Group has once again identified conventionally grown grapes as one of most problematic fruits and vegetables in terms of pesticide residues. There's new evidence that pesticide residues can be successfully avoided by purchase of certified organic grapes. In a recent study of 99 vineyards in the Aegean Sea area of the Mediterranean, pesticide residues were found on conventionally grown table grapes, but were determined to be undetectable on grapes that had been organically grown. This new evidence adds to our confidence about the added health benefits of selecting organically grown grapes.

WHFOODS RECOMMENDATIONS

In our Healthiest Way of Eating Plan, we encourage the consumption of 5-10 servings of fruits-plus-vegetables (combined) eat day. We believe that the balance between fruits and vegetables can vary from day to day, depending upon personal health factors, personal taste preferences, and optimal combining of foods in recipes as well as meals. We recognize that our recommendation calls for a more generous amount of fruits and vegetables than the amount recommended by the Centers for Disease Control (CDC) at the U.S. Department of Health and Human Services (DHHS). The CDC recommends between 1.5-2.5 cups of fruit and 2.5-4.0 cups of vegetables per day, as well as a target goal of at least 5 fruit-plus-vegetable servings (combined) per day. We recommend that you set your fruit goals higher than these CDC amounts. Based on the scientific research, we believe it's going to take closer to 3 fruit servings per day (consisting of one cup's worth of fruit per serving, or 3 cups total per day) to provide you with optimum health benefits.

With respect to berries (and remembering that grapes are included among the berry fruits), we recommend that you include berries at least 3-4 times per week within your fruit servings. In several of our sample meal plans, we include berries on a daily basis! It would definitely not be a mistake for you to include a serving of either grapes, raspberries, blueberries, strawberries, cranberries or other berries in your daily meal plan! When you're including grapes among your daily fruit servings, you should treat one cup as the equivalent of approximately 15-20 grapes. In practice, what this means is that on any given day, if you decide that you would like to consume all 3 of your fruit servings from this delicious food, you can feel confident enjoying up to 45-60 grapes! On another day, if you are primarily in the mood for other kinds of fresh fruit, there is still plenty of room within this 3-serving range to include some fresh grapes on a salad, or enjoy a small cluster along with a snack.

Grapes, fresh 1.00 cup (151.00 grams)
Calories: 104
GI: low
NutrientDRI/DV
vitamin K25%
copper21%
vitamin B28%

This chart graphically details the %DV that a serving of Grapes provides for each of the nutrients of which it is a good, very good, or excellent source according to our Food Rating System. Additional information about the amount of these nutrients provided by Grapes can be found in theFood Rating System Chart. A link that takes you to the In-Depth Nutritional Profile for Grapes, featuring information over 80 nutrients.

HEALTH BENEFITS

Few fruits have garnered as much attention in the health research literature as grapes. Part of the reason may be their widespread presence in diets worldwide. With the exception of Antarctica, grapes are cultivated on all of the earth's continents, and researchers from many different countries have been especially interested in this food. But an even greater part of the reason involves the amazing nutrient composition of grapes themselves. Every year, it seems like the list of health-supportive grape nutrients grows longer, and it can be challenging just to keep up with the many phytonutrients provided by this popular food.

Organized according to science-based categories, the list below will give you a general idea of the phytonutrient richness of grapes. While a single grape variety is unlikely to contain all of the phytonutrients listed below, grapes as a group have been shown to provide us with the following health-supportive nutrients:

- Stilbenes
  — resveratrol
  — piceatannol
  — pterostilbene
- Flavanols
  — catechins
  — epicatechins
  — procyanidins
  — proanthocyanidins
  — viniferones
- Flavonols
  — quercetin
  — kaempferol
  — myricetin
  — isorhamnetin
- Phenolic Acids
  — caffeic acid
  — coumaric acid
  — ferulic acid
  — gallic acid
- Carotenoids
  — beta-carotene
  — lutein
  — zeaxanthin

In addition to the above-listed nutrients, grapes have also been shown to contain the hormone and antioxidant melatonin as well as unique oligopeptides (small protein-like molecules) that have anti-bacterial and other properties.

With their overwhelming number of health-supportive phytonutrients, it is not surprising that grapes have been shown to provide many of our body systems with predictable benefits. Areas of benefit in grape research include the cardiovascular system, respiratory system, immune system, inflammatory system, blood sugar regulating system, and nervous system. Another area of special benefit is cancer prevention, with risk of breast, prostate, and colon cancer emerging as the most likely areas of grape anti-cancer benefits. The following paragraphs will give you a closer look at some key areas of health research on grapes.

ANTIOXIDANT BENEFITS

The wealth of antioxidant nutrients in grapes is somewhat startling! In addition to providing us with conventional antioxidant nutrient like vitamin C and manganese, grapes are filled with antioxidant phytonutrients that range from common carotenoids like beta-carotene to unusual stilbenes like resveratrol, and the total number of different antioxidant nutrients in grapes runs well into the hundreds.

(Even the hormone melatonin has been identified in grapes and is known to act as an antioxidant provided by this food.) It's important to note that the seed and the skin contain the richest concentration of antioxidants. It's very rare to find a higher concentration of an antioxidant in the fleshy part of the grape than is present in the seed or skin.

For this reason, most of the health research on grape antioxidants has not been conducted on whole grapes. Instead, this research has been conducted on grape skin, grape skin extract, grape seed, grape seed extract, or on grape extracts that contain skin and seed and flesh. As a general rule, the flesh of the grape contains approximately 1/20th-1/100th of the total antioxidant capacity of the seed or the skin.

The greater concentration of antioxidants in the skin and seed of grapes does not mean that we don't benefit from eating the whole grape, including the flesh! But it does mean that we need to treat grape studies as a whole as most likely reflecting stronger short-term antioxidant benefits than would be associated with short-term intake of whole grapes.

Research on antioxidant benefits provided by grapes or grape components includes the following findings. Grapes and grape components can:

- help prevent certain oxygen-related enzymes from becoming overactive.
- These enzymes include xanthine oxidase and catalase.
- increase our blood levels of glutathione (a critical antioxidant nutrient) and also increase the ratio of reduced-to-oxidized glutathione (one important measure of antioxidant capacity).
- help protect cell membranes from free radical damage.
- lower levels of oxygen reactive molecules in our blood.
- reduce oxidation of fat (lipid peroxidation).
- lower biomarkers of oxidative stress.

ANTI-INFLAMMATORY BENEFITS

Along with their strong antioxidant support, grapes provide us with equally strong anti-inflammatory benefits. Once again, research studies in this area have seldom involved dietary intake of whole grapes, but rather supplemental intake of grape components or grape extracts.

Still, we have every reason to believe that these same anti-inflammatory benefits are offered by whole, fresh grapes, perhaps just not to the same extent in a short-term situation (like the few weeks or months that characterize most research studies). It's important to remember that we can enjoy grapes over a lifetime!

Grapes have been determined to lower our risk of excessive and unwanted inflammation in a variety of ways. Many pro-inflammatory messaging molecules can have their activity level reduced by grape intake. These molecules include interleukin 6 (IL-6), interleukin 1-beta (IL-1B), and tumor necrosis factor alpha (TNF-alpha). Overproduction of the pro-inflammatory enzymes cyclo-oxygenase 1 and 2 (COX-1 and COX-2) is also less likely following intake of grape components.

CARDIOVASCULAR BENEFITS

No body system is better situated to reap the benefit of antioxidants and anti-inflammatory molecules in grapes than the cardiovascular system. All cells in our blood need protection from potential oxygen damage (especially in our arteries where oxygen concentration in our blood is especially high). Our blood vessel linings also need strong antioxidant support. Chronic inflammation in our cardiovascular system is also a primary concern for many types of cardiovascular disease, and optimal regulation of inflammatory system activity is especially important in lowering our risk of atherosclerosis and other conditions.

The list of cardio benefits provided by grapes and grape components is perhaps the most impressive of all grape benefits. It's also one of the reasons that consumption of grapes in the form of red wine has been regarded by some researchers as a key for understanding "the French Paradox." The French Paradox refers to research observations about heart health in the French population in relationship to their saturated fat intake. Despite eating fairly large amounts of saturated fat in their overall diet, the French population as a whole has been observed to have much lower levels of heart disease than would be expected with high saturated fat intake.

One of the reasons might be the anti-inflammatory (and antioxidant) support provided to their cardiovascular system on a regular basis by red wine. The idea that red wine (from red grapes) could help explain the French Paradox is just one more reason for us to consider grapes as a great addition to a heart healthy diet.

All of the following cardio benefits have been demonstrated in research studies on grapes and grape components:

- better blood pressure regulation, including blood pressure reduction if high
- better total cholesterol regulation, including total cholesterol reduction if high
- reduced LDL cholesterol levels
- reduced LDL oxidation
- reduced levels of reactive oxygen molecules in the blood
- reduced likelihood of cell adhesion to the blood vessel walls
- less clumping together of platelet cells, when inappropriate
- enhanced release of nitric oxide from endothelial cells lining the blood vessel walls in situations where vasodilation is needed
- better inflammatory regulation in the blood
- increase levels of glutathione in the blood

BLOOD SUGAR BENEFITS

In terms of blood sugar regulation, not all fruits are created equal. Watermelon, for example, has a relatively high glycemic index (GI) value in the range of 70-75, and is not considered to be a food that can be freely eaten by persons having difficulty with blood sugar balance. Grapes, on the other hand, have long been classified as a low glycemic index (GI) food, with GI values ranging between 43-53. In the case of grapes, recent studies have also shown that the low GI value of grapes is a good indicator of this fruit's blood sugar benefits. Studies have now connected grape intake to better blood sugar balance, better insulin regulation, and increased insulin sensitivity. We suspect that the strong phytonutrient content of grapes plays a key role in providing these blood sugar-related benefits.

ANTI-AGING AND LONGEVITY BENEFITS

Several grape phytonutrients may play a role in longevity and may provide us with anti-aging benefits. Best-studied in this area of health benefits is resveratrol (a stilbene phytonutrient presently mostly in grape skins, but also in grape seeds and grape flesh).

Resveratrol has recently been shown to increase expression of three genes all related to longevity. (These three genes are SirT1s, Fox0s, and PBEFs.) Interestingly, some researchers have shown a parallel between activation of these longevity genes by resveratrol and activation by calorie-restricted diets. (In aging and longevity research, our ability to get optimal nutrition for the fewest possible amount of calories is related to our longevity, and the more we can decrease our calories while staying optimally nourished, the better our chances of healthy aging and longevity.)

COGNITIVE BENEFITS

Several recent studies on grape extract intake by animals, as well as grape juice intake by humans, suggest that grapes may provide us with some important cognitive benefits. For example, daily consumption of Concord grape juice in a 1-2 cup amount over a period of several months has been shown to improve the scores of study participants on the California Verbal Learning Test. Other studies on animals have shown that excessive accumulation of reactive oxygen species (ROS) in the brain can be prevented with intake of grape extracts, as can excessive accumulation of beta-amyloid protein in the hippocampus region of the brain. Synthesis of pro-inflammatory messaging molecules in the brain (including IL-6, IL-1B, and TNF-alpha) has also been shown to be reduced by intake of grape extracts. While large-scale human studies are needed to confirm these potential benefits, we expect that grape benefits for our cognitive health will be confirmed in future research.

ANTI-MICROBIAL BENEFITS

Numerous grape phytonutrients have been shown to have anti-microbial properties. These phytonutrients range from common flavonoids like quercetin to less common stilbenes like piceatannol and resveratrol. Recent studies have determined that grapes may also contain unique sets of oligopeptides (short protein-like molecules) that have anti-microbial properties. Exactly how we benefit from these anti-microbial substances in grapes is not yet known. But researchers have begun to speculate about their possible role in helping us prevent microbe-related problems like food-borne illness. While research in this area is clearly in its early stages, it will be interesting to see whether diets that are rich in grapes (or grape products like red wine) turn out to be associated with reduced risk of microbe-related problems like food-borne illness.

ANTI-CANCER BENEFITS

The antioxidant and anti-inflammatory properties of grapes make them a natural for protection against cancer because chronic oxidative stress and chronic inflammation can be key factors in the development of cancer. If our cells get overwhelmed by oxidative stress (damage to cell structure and cell function by overly reactive oxygen-containing molecules) and chronic excessive inflammation, our risk of cell cancer is increased. By providing us with rich supplies of antioxidant and anti-inflammatory nutrients, grapes can help us avoid this dangerous combination of chronic oxidative stress and chronic inflammation.

Research on grapes, grape components and cancer has focused on three cancer types: breast cancer, colon cancer, and prostate cancer. In the case of breast cancer, the grape phytonutrients receiving the most research attention have been the catechins, proanthocyanidins, and stilbenes (especially resveratrol). For colon cancer, more research attention has been focused on grape flavonoids and grape anthocyanins. (This research tendency may mean that dark purple, red, and black grapes could end up being better choices for colon cancer prevention than green grapes, even though the jury is still out in this area.) One interesting development in the research on grapes and colon cancer prevention has involved studies on GADF, or grape antioxidant dietary fiber. While GADF is a proprietary product that may eventually be sold in the marketplace as a dietary supplement, the idea that grape fiber and grape antioxidants combine to provide our colon with special support makes perfect sense. Fiber is greatly needed for a healthy colon, and grapes provide us with approximately 1 gram of fiber in every 60 calories. Antioxidants are also needed, and grapes come through strong in the antioxidant category. This antioxidant-plus-fiber combination may be one of the reasons that colon cancer prevention has jumped out in health research on grapes. In the area of prostate cancer, it is the stilbenes (including resveratrol) have been the major focus in research on the anti-cancer properties of grapes.

DESCRIPTION

While table, wine, and raisin grapes come from the same family (*Vitaceae*) and genus (*Vitis*) of plant, there are about 60 different species of this plant type, and within these 60 species, there are literally thousands of grape varieties. (In the country of Italy, for example, more than 1,000 varieties of wine grapes are found in hillside vineyards.) Table grape varieties are most often larger in size, have been propagated to be seedless, and have relatively thin skins. Wine grapes are usually smaller in size, contain seeds, and have relatively thick skins. (Among several different types of benefits, the thicker skin of the wine grapes helps provide the wine with a richer aroma.) All types of grapes come in a variety of colors. While green, red, and black are the most commonly consumed color varieties in the U.S., grape colors also include amber/yellow, blue black, crimson, pink, and purple. (Although you may also hear the color "white" being used to refer to grapes, "white grapes" are actually green in color.)

While we often think about the Mediterranean as the home for all grapes, this amazing food is actually native to many parts of the world, including regions in Asia, Africa, and North America. From a botanical standpoint, one of the most widely cultivated table and wine grapes is *Vitis vinifera* (Common Grape Vine), and several thousand varieties of this grape exist across Europe. In the United States, native species of *Vitis* grapes include *Vitis labrusca* (which includes the Concord grape) and *Vitis rotundifolia* (which includes the muscadine grapes). In Asia, one native grape that is widely-cultivated is *Vitis amurensis* (the Amur grape). From a commercial standpoint, grape varieties from around the world have now been hybridized to produce unique blends of flavor, texture, and environmental suitability.

There has been a good bit of confusion—especially on the internet—about seedless grapes and their production. Many websites mistakenly assume that seedless grapes have been genetically modified. While agricultural researchers are definitely exploring genetically modified grape varieties and while genetically engineered (GE) grapes do exist, they are very rare in the marketplace. Some seedless varieties of grapes are the result of natural mutations, and these varieties can be vegetatively propagated to allow for commercial production. Other grape varieties will produce seedless fruit if pollination is withheld. Still other varieties of seedless grapes can be produced by crossbreeding or grafting. None of these methods involves direct manipulation of the grape plant's genetic material. With many foods, purchase of a certified organic version is one of your best ways to avoid a genetically modified version. In the case of grapes, even though we strongly recommend the organic version for a variety of reasons, avoidance of GE is not one of them. You're unlikely to get GE grapes even if you are purchasing conventionally grown varieties.

RAISINS

Raisins are made from dehydrating grapes in a process that either involves the heat of the sun or a mechanical process of oven drying. Among the most popular types of raisins are Sultana, Malaga, Monukka, Zante Currant, Muscat, and Thompson seedless.

While raisins can be a good addition to a Healthiest Way of Eating, it's important not to see them as a substitute for grapes. That's because dehydration not only reduces the grapes' water content but also increases their concentration of sugar and calories. With these changes, the grapes lose their nutrient richness: in every ounce of raisins, you end up getting four times the amount of sugar and calories that you would be getting in an ounce of grapes, even though you aren't getting any more vitamins and minerals. Additionally, even though raisins are made from grapes, it's not possible to generalize the health benefits shown in research studies on grapes and their components to raisins.

That said, raisins can make a healthful contribution to snacks, cereals, salads, and other recipes. Eating them in moderation is best, given that one-quarter cup contains over 100 calories and the glycemic index rating for raisins is medium rather than low. By comparison, you get to enjoy one full cup of grapes for the same number of calories, with grapes qualifying as a low glycemic index food.

HISTORY

Grapes have a long and abundant history. While they've grown wild since prehistoric times, evidence suggests they were cultivated in Asia as early as 5000 BC. The grape also played a role in numerous biblical stories, being referred to as the "fruit of the vine." Grapes were also pictured in hieroglyphics in ancient Egyptian burial tombs. During the ancient Greek and Roman civilizations, grapes were revered for their use in winemaking. They were planted in the Rhine Valley in Germany, a place of notable wine production, in the 2nd century AD. Around this time, over 90 varieties of grapes were already known.

Even though grapes are known to have grown wild on many continents - including Europe, Asia, Africa, and North America - travel and exploration (together with a cultural fondness for this food and its wines) led to transport of native grapes across the world. For example, as European travelers explored the globe, they brought native European ("old World") grapes with them. Grapes were planted in the United States as early as the 17th century at Spanish missions in the Southwest. They also found a lasting home in the central valley of California where climate, and the relative absence of grape-preying insects, helped support their production.

In the United States, well over 90% of all commercially grown table grapes come from California, where over 700,000 acres of grapes (including table, wine, and raisin grapes) are in cultivation. Nearly 2 billion pounds of table grapes come from these 700,000 acres each year! Early in the year, from January to April, U.S. consumers depend upon imported grapes, including imports from South American countries like Chile. During May and June, imported Mexican grapes are widely available. After June, California grape production takes over the commercial marketplace in the U.S.

Worldwide, about 30,000 square miles of land are planted with grapes, resulting in the annual production of 150 trillion pounds of grapes. Grape production in five countries—Italy, China, Spain, and France—accounts for about half of all commercial production, and about one-third of all world vineyards are found in the three countries of Italy, Spain, and France. Other important grape-producing countries include Turkey, Chile, Argentina, Iran, South Africa, and Australia.

HOW TO SELECT AND STORE

For the best tasting grapes with high concentrations of antioxidants select those that are fully ripe. Fully ripened grapes are plump and free from wrinkles. They should be intact, firmly attached to a healthy looking stem, and not leaking juice. The area around the attachment should have the same color as the rest of the grape.

One way to help predict the sweetness of grapes is by their color: green grapes are medium sweet, red grapes are very sweet, and blue-black grapes are the least sweet. Green grapes should have a slight yellowish hue, red grapes should be mostly red, while purple and blue-black grapes should be deep and rich in color.

Since grapes tend to spoil and ferment at room temperature, they should always be stored in the refrigerator. Loosely wrap unwashed grapes in a paper towel and place them in an airtight container or plastic bag. This way, they'll keep fresh in the refrigerator for up to 5 days.

While freezing detracts from some of their flavor, frozen grapes are a wonderful snack and particularly intriguing to children. To freeze grapes, wash and pat them dry, then arrange in a single layer on a cookie sheet and place in freezer. (Patting the grapes dry is important, because excess moisture can cause the grapes to stick together and become one large frozen fruit clump.) Once frozen, transfer grapes to a heavy plastic bag and return them to the freezer.

TIPS FOR PREPARING AND COOKING

Tips for Preparing Grapes

Grapes should be washed under cold running water right before consuming or using in a recipe. After washing, either drain the grapes in a colander or gently pat them dry. If you are not going to consume the whole bunch at one time, use scissors to separate small clusters of grapes from the stem instead of removing individual grapes. This will help keep the remaining grapes fresher by preventing the stem from drying out.

While some recipes call for peeled grapes, evaluate the recipe to see whether including the skin would actually greatly change the taste and texture, since the grape skin contains many of the fruit's vital nutrients. If retention of the skin isn't objectionable for recipe reasons, we recommend that you always include the skin on account of its nutrient richness. If you decide to use peeled grapes, it's easier to use the American varieties since their skin more readily pulls away from the pulp.

Healthiest Way of Preparing Grapes

Grapes retain their maximum amount of nutrients and their maximum taste when they are enjoyed fresh and not prepared in a cooked recipe. Cooking temperatures used in baking of 350°F/175°C can damage some of the unique and delicate phytonutrients found in this amazing fruit.

HOW TO ENJOY

A Few Quick Serving Ideas

- Grapes are a wonderful addition to any fruit salad. For an enhanced visual effect, consider using a few different varieties of grapes.
- Give your curries a fruity punch by including fresh grapes in the recipe.
- Add grapes to mixed green salads.

- Grapes are great served with cheese as a snack or within a green salad.

WHFoods Recipes That Feature Grapes

- Yogurt with Fruit
- 10-Minute Fruit & Cheese Salad
- Grape and Arugula Salad
- Grape and Melon Salad
- 5-Minute Grapes in Honey-Lemon Sauce
- Ginger Yogurt with Fruit

INDIVIDUAL CONCERNS

If you are drinking grape juice for health benefits, avoid products labeled as grape "drinks." This is often an imitation high-sugar product with little real grape juice.

Grapes and Pesticide Residues

Virtually all municipal drinking water in the United States contains pesticide residues, and with the exception of organic foods, so do the majority of foods in the U.S. food supply. Even though pesticides are present in food at very small trace levels, their negative impact on health is well documented.

The liver's ability to process other toxins, the cells' ability to produce energy, and the nerves' ability to send messages can all be compromised by pesticide exposure. According to the Environmental Working Group's 2014 report "Shopper's Guide to Pesticides," conventionally grown grapes are among the top 12 fruits and vegetables on which pesticide residues have been most frequently found. Therefore, individuals wanting to avoid pesticide-associated health risks may want to avoid consumption of grapes unless they are grown organically.

Nutritional Profile

Grapes are an outstanding source of phytonutrients, especially phenols and polyphenols. The phytonutrients in grapes can differ not only due to growing conditions but also from species to species and variety to variety. Taken as a group, grapes offer phytonutrients in 5 basic categories: flavanols and flavonols (both flavonoids); phenolic acids; stilbenes; and carotneoids. Grape flavonoids include catechins, epicatechins, procyanidins, quercetin, and petunidin. Grape phenolic acids include coumaric acid, caffeic acid, ferulic acid, and gallic acid. Stilbenes in grapes include resveratrol, piceatannol, and pterostilbene. The carotenoids beta-carotene, lutein, and zeaxanthin are also provided in valuable amounts in many varieties of grape. Grapes are a very good source of vitamin K and copper as well as a good source of vitamin B2.

In-Depth Nutritional Profile

In addition to the nutrients highlighted in our ratings chart, an in-depth nutritional profile for Grapesis also available. This profile includes information on a full array of nutrients, including carbohydrates, sugar, soluble and insoluble fiber, sodium, vitamins, minerals, fatty acids, amino acids and more.

INTRODUCTION TO FOOD RATING SYSTEM CHART

In order to better help you identify foods that feature a high concentration of nutrients for the calories they contain, we created a Food Rating System. This system allows us to highlight the foods that are especially rich in particular nutrients. The following chart shows the nutrients for which this food is either an excellent, very good, or good source (below the chart you will find a table that explains these qualifications).

If a nutrient is not listed in the chart, it does not necessarily mean that the food doesn't contain it. It simply means that the nutrient is not provided in a sufficient amount or concentration to meet our rating criteria. (To view this food's in-depth nutritional profile that includes values for dozens of nutrients - not just the ones rated as excellent, very good, or good - please use the link below the chart.) To read this chart accurately, you'll need to glance up in the top left corner where you will find the name of the food and the serving size we used to calculate the food's nutrient composition.

This serving size will tell you how much of the food you need to eat to obtain the amount of nutrients found in the chart. Now, returning to the chart itself, you can look next to the nutrient name in order to find the nutrient amount it offers, the percent Daily Value (DV%) that this amount represents, the nutrient density that we calculated for this food and nutrient, and the rating we established in our rating system. For most of our nutrient ratings, we adopted the government standards for food labeling that are found in the U.S. Food and Drug Administration's "Reference Values for Nutrition Labeling." Read more background information and details of our rating system.

Grapes, fresh 1.00 cup 151.00 grams Calories: 104 GI: low				
Nutrient	**Amount**	**DRI/DV (%)**	**Nutrient Density**	**World's Healthiest Foods Rating**
vitamin K	22.05 mcg	25	4.2	very good
copper	0.19 mg	21	3.6	very good
vitamin B2	0.11 mg	8	1.5	good

World's Healthiest Foods Rating	**Rule**
excellent	DRI/DV>=75% OR Density>=7.6 AND DRI/DV>=10%
very good	DRI/DV>=50% OR Density>=3.4 AND DRI/DV>=5%
good	DRI/DV>=25% OR Density>=1.5 AND DRI/DV>=2.5%

GRAPES NUTRITION FACTS

One of the popular among everyday fruits, grapes are widely viewed in many cultures as "the queen of fruits" since earlier times. These tiny berries of Europe and Mediterranean regions are the storehouse of numerous health promoting phyto-nutrients such as poly-phenolic antioxidants, vitamins, and minerals. No wonder why so many of us often include them in our diet, be it in the form of fresh table fruits, dry fruits, juice, or simply in salads!

Botanically, they are small, round berries growing in clusters on a perennial and deciduous woody vine in the genus,*Vitis*. Today, they are widely cultivated under supervised orchards and vinyards all around the world.

In structure, each grape berry features semi-translucent flesh encased inside a smooth, thin skin. Some varieties contain edible seeds, while others are seedless. The color of the berry is because of the presence of poly-phenolic pigments in them. Red or purple berries are rich in anthocyanins while white-green berries contain more of tannins,especially, catechin. Interestingly, these antioxidant compounds are densely concentrated in the skin and seeds!

The three main species of grapes grown around the world are; European (*Vitis vinifera*), North American (*Vitis labrusca,* and *Vitis rotundifolia*), and French hybrids.

- Some of popular green cultivars are Thompson seedless, sugarone, and calmeria.
- Red varieties include emperor, red globe, cardinal, and flame seedless.
- Concord and zinfandel are some of flavorful blue-black cultivars.

Commercially, many cultivars of grapes are grown for different purposes either to be eaten as table fruit, fresh or dried (raisin, currant, sultana) *or* in wine production.

HEALTH BENEFITS OF GRAPES

- Grapes are rich in polyphenolic phytochemical compound resveratrol. Resveratrol is one of the powerful anti-oxidant which has been found to play a protective role against cancers of colon and prostate, coronary heart disease (CHD), degenerative nerve disease, Alzheimer's disease and viral/ fungal infections.
- Resveratrolreduces stroke risk by altering the molecular mechanisms inside the blood vessels. It does so, firstly by reducing the susceptibility of blood vessel through decreased activity of angiotensin (a systemic hormone causing blood vessel constriction that would otherwise elevate blood pressure) and secondly, through increased production of vasodilator substance, nitric oxide (a beneficial compound that causes relaxation of blood vessels).
- Anthocyanins are another class of polyphenolic anti-oxidants present abundantly in the red grapes. These phyto-chemicals have been found

to have an anti-allergic, anti-inflammatory, anti-microbial, as well as anti-cancer activity.

- Catechins, a type of flavonoid tannin group of anti-oxidants, found in the white/green varieties have also shown to possess these health-protective functions.
- In addition, the berries are very low in calories. 100 g fresh grapes just provide 69 calories but zero cholesterol levels.
- Grapes are rich source of micronutrient minerals like copper, iron and manganese. Copper and manganese are an essential co-factor of antioxidant enzyme, superoxide dismutase. Iron is especially concentrated in raisins. In addition, 100 g of fresh grapes contain about 191 mg of health benefiting electrolyte, potassium.
- They are an also good source of vitamin-C, vitamin A, vitamin K, carotenes, B-complex vitamins such as pyridoxine, riboflavin, and thiamin.

SELECTION AND STORAGE

Fresh grapes can be available all around the seasons. In the stores, choose those that feature plump in consistency, free from surface wrinkles with intact skin, without any cuts or cracks or leaking juice, and firmly attached to a healthy-looking green stem. Lift up the whole bunch in the air and shake gently; loose berries, if any falls off easily.

Buy exactly ripe berries; green grapes should feature slight yellowish hue; red types should be mostly pinkish-red, while purple and blue-black types should be deep and rich in color.

Since grapes tends to spoil early and ferment at room temperature, they should always be stored inside the refrigerator. Loosely wrap washed berries in a paper towel and place them in a zip pouch bag set at high relative humidity. This way, they will keep fresh inside the refrigerator for few days.

PREPARATION AND SERVING METHOD

Just before eating, wash the whole bunch to remove any pesticide residues and dirt by rinsing in cold-water for couple of minutes. Place then in fresh cold water and gently swish them around few times. Pat dry using a soft cloth.

If you are not going to consume the whole bunch at one go, then separate it into small clusters using scissors. This way, you can keep the remaining berries fresher by preventing their stem from drying out.

Although, its seeds are rich in nutrients, seedless grapes make recipes much more pleasant to eat.

Here are some serving tips;

- Enjoy them as they are,without any seasonings/additions.
- Add seedless berries complement in any fruit/vegetable salads with

peach, pear, tomato, lettuce, apricot ,berries, currants, apple, pineapple etc.

- Dry grapes (raisin, currant, and sultana) are being used in confectionary, puddings, cakes, muffins, bread, etc.
- The fruit can also be used by food industry for the preparation of jams, jellies, juice and wine.

HEALTH BENEFITS OF GRAPES

The health benefits of grapes include their ability to treat constipation, indigestion, fatigue, kidney disorders, macular degeneration and the prevention of cataracts. Grapes, one of the most popular and delicious fruits, are rich sources of vitamins A, C, B6 and folate in addition to essential minerals like potassium, calcium, iron, phosphorus, magnesium and selenium.

Grapes contain flavonoids that are very powerful antioxidants, which can reduce the damage caused by free radicals and slow down aging.Grapes, due to their high nutrient content, play an important role in ensuring a healthy and active life.

HEALTH BENEFITS OF GRAPES

Some of the health benefits of grapes include the following:

Asthma: Due to their well-known therapeutic value, grapes can be used as a cure for asthma. In addition to that, the hydrating power of grapes is also high, which increases the moisture present in the lungs and reduces asthmatic events.

Bone Health: Grapes are a wonderful source of micro-nutrients like copper, iron, and manganese, all of which are important in the formation and strength of the bones. Adding grapes into your diet on a regular basis can prevent the onset of age-related conditions like osteoporosis. Manganese is an extremely important element in the body, which aids in everything from protein metabolism, collagen formation, and nervous system functioning.

Heart diseases: Grapes increase the nitric oxide levels in the blood, which prevents blood clots. Therefore, grapes are an effective way to reduce the chances of heart attacks. Additionally, the antioxidants present in grapes prevent the oxidation of LDL cholesterol, which blocks the blood vessels and is a main contributor to various coronary conditions. Grapes also have high numbers of flavonoids, which are what give grapes their color, but flavonoids are also very powerful antioxidants.

The two main types in grapes are resveratrol and quercetin, and these two compounds negate the effects of free radicals that threaten the body and stimulate LDL cholesterol's harmful effects on arteries. Also, these two antioxidant flavonoids act as a clean-up crew to reduce platelet clumping and filter toxins out of the blood.

Migraine: Ripe grape juice is an important home remedy for curing migraines. It should be drunk early in the morning, without mixing additional water. Ironically, drinking red wine is often considered a cause of migraines, but grape juice and grape seed extract is considered a solution for the problem. There are many causes behind migraines, and it can be difficult to pinpoint the culprit, since they include chemical imbalances, lack of sleep, changes in weather, or dietary deficiencies. Alcohol in general causes migraines, but grapes have so many antioxidants that they can cause and cure the same illness!

Constipation: Grapes are very effective in overcoming and eliminating constipation. They are classified as a laxative food, because they contain organic acid, sugar and cellulose. They also relieve chronic constipation by toning up intestinal muscles and the stomach. Grapes are high in insoluble fiber, meaning that it remains intact as it moves through the digestive tract. It builds up bulk, which promotes the formation and excretion of healthy stool, so grapes can help make you much more regular. However, if you suffer from loos stool or diarrhea, grapes should not be used as a way to regulate your system. Insoluble fiber doesn't soak up water to dry out loose stool, and grapes don't have a high level of *soluble* fiber.

Indigestion: Grapes play an important role in dyspepsia. They relieve heat and cure indigestion and irritation in the stomach. They are also preferred over other digestive aids because they are considered a "light food".

Fatigue: Light and white grape juice supplements the iron content in the body and prevents fatigue. Anemia is a real problem for many people, and eating grapes can help keep your iron and mineral levels balanced in the body. A lack of iron can make you sluggish, and your mind also doesn't work as quickly, since iron is an essential mineral that impact a number of bodily functions. However, dark grape juice might not give an iron boost and may actually decrease iron levels. Drinking grape juice also provides a nearly instant energy boost.

Diabetes: Recent research suggests that grape skin extract (GSE) exerts a novel inhibitory activity on hyperglycemia and may help in diabetes management.

Dental Care: According to a recent study, red wine and grape seed extract can potentially help prevent cavities.

Kidney disorders: Grapes can substantially reduce the acidity of uric acid and they also help to eliminate acid from the system, thereby reducing the stress andpressure on the kidneys. Since grapes have a high water content, they induce urination, which also helps to eliminate the uric acid still present in the body after its acidity is reduced. Grapes have a very cleansing effect on the body, and the antioxidants present benefit all of the body's systems in peripheral ways.

Blood cholesterol: Grapes contain a compound called pterostilbene, which has the capacity to lower a person's cholesterol levels. Pterostilbene is closely related to resveratrol, the beneficial antioxidant and coloring flavonoid that is also found in grapes, and early research has shown that it has anti-cancer qualities as well as having a great impact on cholesterol levels. It is also thought to have preventative powers over cognitive decline. Furthermore, the saponins present in the skin of grapes can also prevent the absorption of cholesterol by binding with it.

Antibacterial activity: Red grapes have strong antibacterial and antiviral properties that can protect you from infections. They display strong antiviral properties against the polio virus and the herpes simplex virus. Studies have also shown that grape juice can tackle bacterial infections in the gut and other systems.

Breast cancer: In a recent study, it has been discovered that purple, Concord grape juice helps in preventing breast cancer. Significant reduction in mammary tumor mass of laboratory rats was seen after they were fed the grape juice in an experimental setting. The antioxidants in grapes are beneficial in the prevention of all types of cancer, but there are some fascinating studies which specifically show grape juice as an effective preventive measure against breast cancer. Certain chemicals in grapes compete for binding sites, and inhibits aromatase from changing androgen to estrogen, which is thought to be a big contributing factor to breast cancer growth in women.

Alzheimer's disease: Resveratrol, a beneficial polyphenol present in grapes, reduces the levels of amyloidal-beta peptides in patients with Alzheimer's disease. Studies suggest that grapes can enhance brain health and delay the onset of degenerative neural diseases. A study published in the British Journal of Nutrition claims that grape juice can even improve the brain function of older people who have already displayed mild cognitive impairment. Further studies are currently under way to determine the exact interaction between grape components and neural systems.

Macular degeneration: Grapes can prevent age-related loss of vision and macular degeneration. Research study conducted at University of Miami suggests that grape-enriched diet supports eye health and may prevent vision-threatening retinal diseases.

Immune System: Grapes are not only packed with flavonoids and minerals, but vitamins as well! The high levels of Vitamin C, K, and A in grapes gives you a healthy boost to many of your organ systems, particularly your immune system, which means less chance of coming down with common colds, as well as more serious health issues.

Prevention of cataracts:Flavonoids present in grapes have antioxidants, which can reduce and fight the damage caused by free radicals. Free radicals cause things like the development of cataracts, as well as cardiovascular diseases, cancer, and various age-related problems. This is somewhat linked to macular degeneration, as they usually occur around the same time in life. Fortunately, antioxidants also have certain anti-aging qualities, which not only help you look younger, but also feel younger by staving off age-related conditions like cataracts.

Cancer prevention properties: Grapes have been found to have strong anti-cancer properties due to the anti-inflammatory effect of resveratrol. It is particularly effective in reducing the chances of colorectal cancer and breast cancer. The anthocyanins and pro-anthocyanidins in grapes have properties of an anti-proliferate and can inhibit the growth of cancer causing agents. Grape juice not only prevents the risk of cancer but also suppresses the growth and propagation of cancer cells. The pigments contained in grapes enhance the overall immunity of the body to a wide range of diseases.

In conclusion, grapes play a pivotal role in preventing a multitude of health disorders and can be used as a home-based remedy for several ailments. Dried grapes, known as raisins, are also extremely nutritious and can help treat many disorders, including constipation, acidosis, anemia, fever, and sexual dysfunction. Raisins can also help people gain weight quickly and protect the health and functional integrity of their eyes.

NUTRITION BENEFITS OF RED GRAPES

People throughout history have consumed grapes for their potential health benefits. In the beginning of the 20th century, a grape diet was thought to prevent or cure cancer and other illnesses. This belief may have stemmed from the findings of Johanna Brandt, a South American dietitian who claimed to have cured her stomach cancer by following the diet.

However, no scientific evidence has proven that eating grapes, either as part of a grape diet or in conjunction with other healthy foods, can prevent or cure cancer or other illness. Despite this fact, red grapes do have several health benefits and make a nutritious snack.

Fig. Red grapes contain a number of essential vitamins and minerals.

NUTRITION FACTS

Red grapes are relatively low in calories and are fat and cholesterol free. A 1-cup serving of red grapes contains only 104 calories, yet still provides 1 g of protein and 1 g of fiber. However, this serving also contains 27 g of total carbohydrates, 23 g of which come from sugar.

VITAMINS

Red grapes are a good source of several vitamins. One serving of red grapes offers 16 mg of vitamin C, which is about 27 percent of the daily value, based on a 2,000-calorie diet. It also contains 22 mcg of vitamin K and 0.4 mg of thiamin, which is about 28 percent and 27 percent of the daily value, respectfully. Red grapes also contain small amounts of vitamin E, vitamin A, vitamin K, pantothenic acid, vitamin B-6, folate, beta carotene and alpha carotene. These nutrients help your body processes function properly, and many vitamins, such as vitamin A, have antioxidant properties which may help reduce the risk for certain help problems.

MINERALS

Red grapes are also rich in several minerals. A 1-cup serving of red grapes contains 288 mg of potassium, 0.2 mg of copper, 0.1 mg of manganese and 0.5 of iron — 10 percent of the daily value of potassium and copper, 5 percent of the daily value of manganese and 1 percent of the daily value of iron. Red grapes also contain small amounts of calcium, magnesium, zinc and selenium.

RESVERATROL

Red grapes contain resveratrol, which is a polyphenol antioxidant. This antioxidant is concentrated in the skins of red grapes. A 1-cup serving of red grapes contains about 160 g of resveratrol, according to the Linus Pauling

Institute. Resveratrol seems to have anti-inflammatory, antioxidant and antiestrogenic properties, and may activate the enzymes of the liver that rid the body of unwanted chemicals, according to the American Cancer Society. It is possible that resveratrol may also play a role in preventing cancer and heart disease, as Brandt believed, and could extend a person's life, but more research needs to be conducted to confirm these theories.

HEALTH BENEFITS OF GRAPES

Grapes are loaded with nutrients and provide many of the same health benefits of wine, without the alcohol.

Phytonutrients, plant chemicals that contain protective, disease-preventing, compounds. are bioactive compounds that are not designated as traditional nutrients, but have been shown to have health benefits in humans – grapes are loaded with them, particularly in the skin.

Resveratrol is a polyphenol antioxidant synthesized by the grape plant for antifungal and defensive properties. Numerous research studies have demonstrated the health benefits of resveratrol. It lowers risk of cardiovascular disease by preventing lipid oxidation and it modulates levels of lipids and lipoproteins in the body. Resveratrol is also a potent antioxidant, which not only reduces oxidation, but also helps to strengthen other antioxidant substances in the body.

The polyphenols found in grapes, have been shown to reduce the risk of heart disease by reducing platelet aggregation which prevents blood clots, protect low density lipoprotein (LDL) or "bad cholesterol" from free radical damage, lower blood pressure, and improve the function of blood vessels.

Grapes are also an excellent source of manganese and a good source of vitamin B6, thiamin, potassium, and vitamin C.

GRAPES - HISTORY + USE

Grapes grow in clusters of 6 to 300 on vines. They range in color from crimson to black to yellow. Grapes were first cultivated as early as 5,000 BC. They are mentioned in many biblical stories where they are referred to as "fruit of the vine".

European travelers spread them all over the world during their travels. They were first brought to the US in the 17th century, where they migrated to the central valley of California and thrived due to the climate and absence of grape-preying insects. Today California produces most of the grapes consumed in the US.

Grapes should be stored in the refrigerator and washed right before eating. To keep them fresh longer, grapes can be frozen by arranging them in a single layer on a cookie sheet and placing them in the freezer. Once they are frozen, they can be stored in a plastic bag. Frozen grapes make a delicious snack, especially on a hot day.

GRAPE TIPS

Grapes can be enjoyed alone as a snack or used in a variety of recipes. Below are some ways to increase the amount of grapes in your diet:

- Serve stewed grapes with meat items
- Add grapes to rice dishes for a touch of sweetness
- Add to a green salad or fruit salad
- Place sliced grapes on your favorite sandwich
- Serve grapes with cheese and crackers

Grape juice is an excellent way to take advantage of the health benefits of grapes without the alcohol content of wine. Studies have shown that juice made from concord grapes have the highest amount of antioxidants when compared to other juice products.

BLACK GRAPES HEALTH BENEFITS

Black grapes have been associated with health and nutrition for centuries and they even appear in ancient literature and works of art. The very word 'grapes' conjures up images of bunches of plump black grapes, the jeweled wine goblets of the ancient Romans, and the splendors of a bygone era. The nutritional value of black grapes however, has stood the test of time and in fact new research only helps to uncover more health benefits of this fruit. Here are some of the black grapes health benefits that deserve a mention:

- Blood Sugar – Black grapes have glycemic index values that range from around 43 to 53and so they are listed as low GI foods. A low glycemic index isn't necessarily indicative of blood sugar benefits but in the case of black grapes, their low GI value is clearly reflected in significant blood sugar benefits.

ADVERTISEMENT

Several studies show that the consumption of grapes, as well as grape juices and extracts promote better blood sugar balance, improved regulation of insulin, and increased insulin sensitivity.

- Cardiac Health – The cardiac health benefits of black grapes are probably most noteworthy, as studies indicate that they can drastically reduce heart disease risks. Research at the University of Michigan Cardiovascular Center suggests that grapes can be used to treat as well as prevent heart diseases. These benefits can be traced to the action of the phytochemicals in grapes that actually trigger a protective process in certain genes, reducing damage to the heart muscles. This is why the effect of black grapes on heart health is considerably greater than the simple blood pressure lowering effect of a diet rich in fruits and vegetables. The consumption of grapes also reduces and regulates cholesterol levels.
- Brain Function – The regular consumption of black grapes improves brain function, enhances memory, and concentration. Studies also seem to suggest that individuals who consume grape juice on a daily basis are less likely to succumb to brain diseases like Alzheimer disease.
- High Flavonoid Content – So, what's so special about flavonoids? Flavonoids don't just lend grapes their great taste, but they also are responsible for most of the health benefits. The color and juiciness of the grapes derive from these nutrients, which are phytochemicals. So, higher the flavonoid content, the darker and juicier the quality of the grapes. Correspondingly, black grapes are believed to be the healthiest.
- Resveratrol – This is one of several grape phytonutrients that is believed to affect longevity, because of its effect in slowing oxidization

of cells and molecules, a property that also offers protection against degenerative conditions like cancer. Grape skins, grape seeds and grape flesh, all have high concentrations of this phytonutrient.

- Quercitin – This is another phytonutrient present in grapes that has been found to demonstrate anti-inflammatory and antioxidant properties. Research also indicates that this particular property could lower the risks of developing a blood clot.

NUTRITIONAL INFORMATION ON BLACK GRAPES

Black grapes aren't just delicious; this tiny fruit is packed with nutrients. Juggling between the various choices of foods to meet those varied nutritional requirements can be quite a challenge, but including just a handful of black grapes in your daily diet can get you a lot closer to your nutritional goals.

The nutritional value of black grapes makes it an extremely valuable health food, with a single serving of around 80 g containing:

- Calories – 60 kcal
- Carbohydrates – 14.0g
- Vitamin C – 15% of the average RDA
- Protein – 1.0g
- Dietary Fiber – 1g

SIDE EFFECTS OF BLACK GRAPES

There are no known black grapes side effects from the moderate consumption of black grapes, or any other variety of grapes, unless you suffer from a grape allergy. Grape seed and grape seed extracts however do have counter indications, so check with your doctor if you suffer from any pre-existing condition or if you are on any medications.

Black Grapes Juice Medicinal Benefits:

With many fruits, the health benefits are best obtained from the consumption of the fruit itself, but with black grapes, the juice too offers medicinal benefits that are just as vast. The juice of black grapes retains the nutritional value of grapes in their raw form, and consequentially offers the same health benefits. Methods of extraction and processing vary however, and the presence of additives and preservatives could compromise the health benefits of the processed product.

Black Dry Grapes Nutrition Benefits:

Black dried grapes are known as raisins and are often added to breakfast cereal and desserts. Black dry grapes nutrition benefits include protection against osteoporosis (due to their boron content) and protection from eye vision loss.

Advantages of Eating Black Seedless Grapes:

According to research in the 2008 "Journal of Environmental Pathology, Toxicology and Oncology", one of the main health benefits of black seedless grapes is their ability to reduce cancer risks because of their polyphenols content. There is little difference in the nutritional value between black grapes with seeds and the seedless varieties.

Black Grapes vs Green Grapes & Red Grapes:

The main differences in grapes based on skin color, such as between black grapes or green grapes, are in the context of their antioxidant properties. Darker colored grapes are found to display higher anti-oxidant activity as compared to lighter varieties. As most of the health benefits in grapes derive from flavonoids and phytonutrients that lend them their color, black grapes are thought to be even more nutritious as compared to most other varieties. Black grapes also derive some of their special health benefits from the presence of nocyanosis, a substance that lends it its distinctive color and tonic properties as well!

3

Major Diseases of Grapes

INTRODUCTION

Rapid, accurate identification of diseases in the vineyard is key to preventing serious outbreaks and losses in yield and quality. However, the presence of a pathogen or disease does not necessarily mean that a treatment is required.

The severity of diseases varies from year to year, depending primarily on weather conditions, the presence of inoculum (history of the disease) and the susceptibility of the vines. This means that a disease can be devastating one year and insignificant the next. The measures to be taken to prevent losses may therefore vary from season to season.

The purpose of this book is to aid in the identification of grape diseases and in pest management decision-making. Accurate identification of pests is critical to an effective management program that provides optimal control while minimizing pesticide use.

LEGEND

Favourable conditions for disease development:

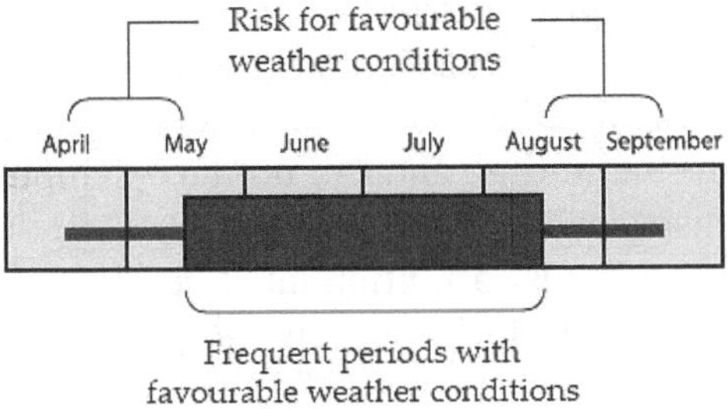

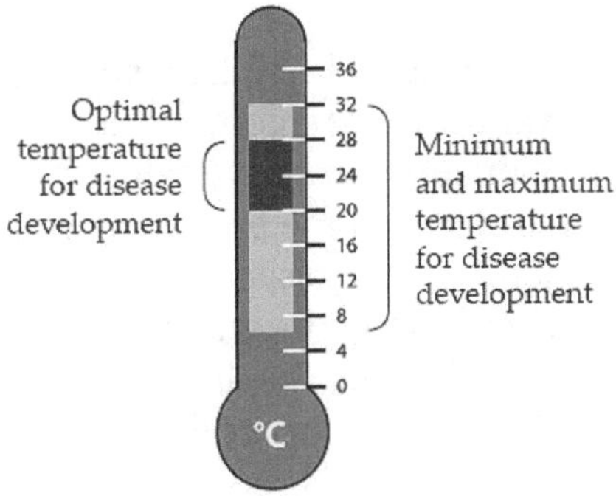

SCOUTING

When scouting a vineyard, bear in mind that there are several factors that can cause damage, including diseases, insects, herbicides, mammals, birds, cultural practices and abiotic factors (e.g., weather). It is important not to draw premature conclusions regarding the cause of a problem and to keep in mind that more than one disease can be present at the same time. Ideally, a vineyard, regardless of whether or not it is in production, should be completely scouted, i.e., all sections and all vines, at least once a week and ideally twice a week, from bud break to harvest.

However this is not generally practical so special attention should be given to susceptible varieties or areas prone to disease, where the first symptoms of the disease generally appear. Well-targeted action taken early in the outbreak will generally result in improved disease control, often with less use of pesticides.

Following harvest, it is important to check the condition of the vines before leaf drop (e.g. downy or powdery mildew sporulation) and after (e.g. anthracnose cankers on the cane) to assess the inoculum present on the leaves and cane at the end of the season. It is also easier to detect diseases that affect the wood of the vines after the leaves have dropped.

Examples of possible confusion during scouting

- Some diseases have similar signs and symptoms.
- A disease can have different signs and symptoms, sometimes depending on the grape variety.
- Several diseases can be present at the same time.
- Several factors, weather conditions, deficiencies, pesticides, can be present at the same time and can produce symptoms that are similar to those of diseases.

During scouting, particular attention must be paid to the following points

1. Clearly identify the block, susceptibility of grape variety, and vine growth stage.
2. Date of appearance of symptoms or damage, weather conditions before and during scouting.
3. Distribution of the damage within the vineyard, e.g., localized, limited to certain grape varieties or to certain parts of the vineyard, such as a depression. It is important to cover all parts of the vineyard (relief, grape variety). Knowledge of the prevailing wind direction is important for spread orientation of wind dispersed diseases like powdery mildew.
4. The distribution of damage on the vine: all leaves, young or old leaves, upper side or underside of leaves, berries.
5. The description of the symptoms or damage: spot, scorch, rot, wilting. Check for steel wire or any other material that could injure plant tissues.

DISTRIBUTION OF SYMPTOMS OR DAMAGE:

- In block: Located near the wooded area, in a depression, presence of water on the ground. Incidence: several vines, less than 1%
- On vine: Primarily at the apex of the stems (Photo 1) White sporulation on the lower leaf surface, less than 25% of the vine affected

Description of Symptoms or Damage: Infected shoot tips curl and are covered with white sporulation (Photo 2) White sporulation can be removed with the fingers

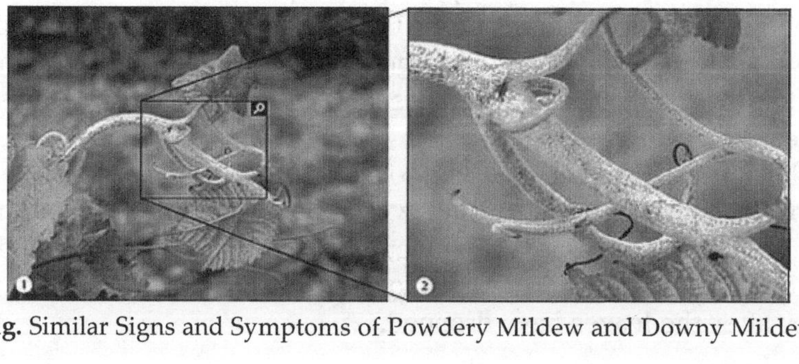

Fig. Similar Signs and Symptoms of Powdery Mildew and Downy Mildew

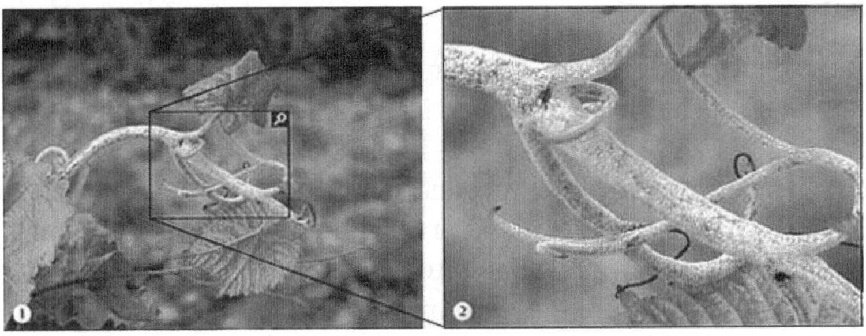

Fig. Sporulating lesions - upper surface (30° angle) powdery appearance

Fig. Sporulating lesions - lower surface (90° angle) downy appearance

MANAGEMENT OF GRAPE DISEASES

PREVENTION

There are a number of factors that must be considered in selecting grape varieties. Planting varieties that are less susceptible to disease is a good way to prevent outbreaks. Orienting rows N-S and using topographical gradients will maximize air and soil drainage. Pruning debris may act as a reservoir for inoculum and should be removed and burned. Working the soil in the spring is a form of cultural control. Destroying and burying infected debris will reduce pathogen populations at the start of the season, therefore reducing the risk of disease.

Canopy management (pruning, shoot positioning, leaf removal) facilitates air circulation, which promotes drying of the leaves and increased penetration of fungicides.

It is also recommended that vineyards be systematically inspected throughout the growing season to monitor the appearance and development of diseases. At harvest and at the end of the season, assess the presence of diseases like downy and powdery mildew or anthracnose on the leaves to estimate the level of inoculum that will be present the following spring.

CONTROL

There are a number of fungicide application strategies. Where minimum risk is desired, fungicides can be applied in keeping with an established schedule rather than on the basis of observations made in the vineyard. This approach is not without its consequences on production costs and the environment. It is also possible to develop rational IPM programs for diseases of grapes. In such cases, the decision to spray is made after evaluating the consequences of this decision on production and the environment. In other words, the decision will be based on factors such as the weather conditions, the stage of development of the vine, the grape variety, parasitic pressure, etc.

MAJOR GRAPE DISEASES

Grapes are a crop that is susceptible to many diseases. However, the degree of susceptibility varies depending on the variety. When no pest management is carried out, damage can generally be severe. Downy mildew (a), powdery mildew (b), grey mold (c),anthracnose (d), black rot (e) and crown gall (f) are the major grape diseases in Eastern Canada. Anthracnose is not a problem in Ontario to date. Downy mildew, powdery mildew, grey mold, black rot and anthracnose are caused by fungi that attack the berries, reducing yield and quality. Crown gall is caused by a bacterium and can kill the plant.

The first step in an effective pest management program is correct identification of the disease. Since the severity of the damage and losses varies from year to year, the pest management program will vary depending on the weather conditions, grape variety and abundance of the pathogen.

Fig. Downy mildew

Fig. Powdery mildew

DOWNY MILDEW

SYMPTOMS

Leaves: Roughly circular yellowish discolorations, called oil spots (a). White down (sporulation of the fungus) (b) , particularly on the lower leaf surface. The spots turn brown with time (c) and severely infected leaves may drop.

Shoots: Infected shoot tips curl ("shepherd's crook") (d) and a white down occurs on the stem (sporulation of the fungus) (e).

Inflorescences: The inflorescences are highly susceptible. During severe infections, they can turn yellow, brown, then dry out completely (f). They may also exhibit the cork-screwing observed in infected shoots.

Berries: White down (sporulation of the fungus) (g). Berries that exhibit sporulation when they are shot-size may shrivel and drop off. Infected berries of red varieties turn colour prematurely while those on white varieties acquire a mottled appearance. Infected berries tend to stay hard when the rest of the cluster starts to soften at veraison.

DOWNY MILDEW SYMPTOMS

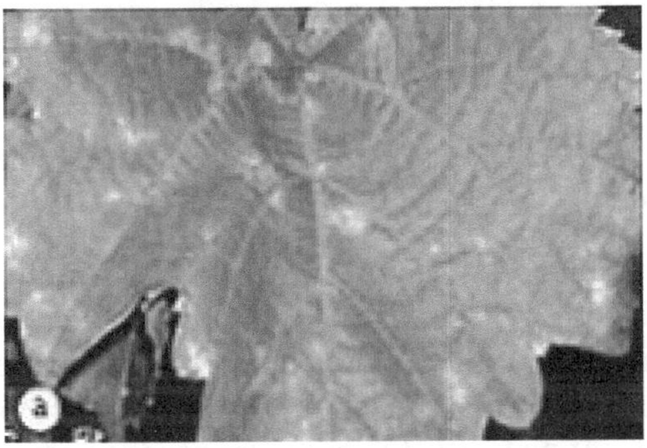

Fig. Oil spots

Fig. White down (sporulation of the fungus)

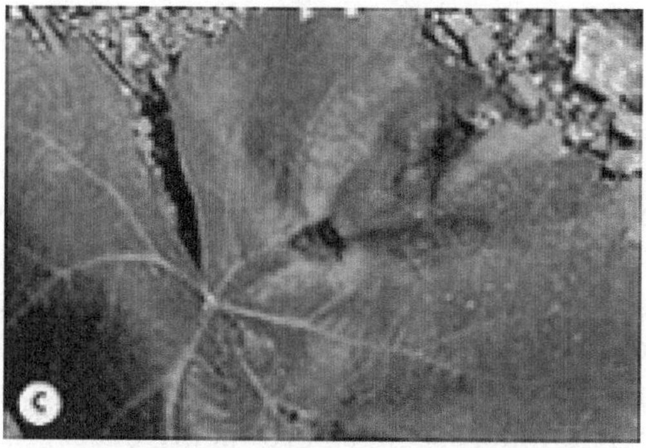

Fig. The spots turn brown with time

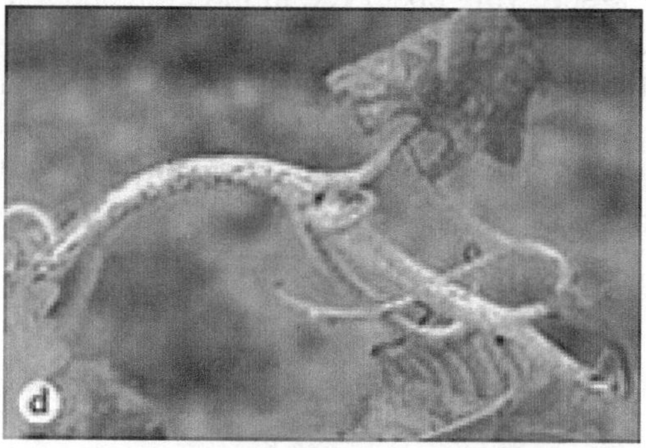

Fig. Infected shoot tips curl

EPIDEMIOLOGY

The causal fungus, *Plasmopara viticola*, is an obligate parasite that overwinters as oospores, or sexual spores, in dead leaves. The quantity of mature oospores in the spring is determined by the amount of precipitation the preceding fall. In the spring, the minimum temperature for the development of the fungus is 11°C. At maturity, the oospores produce new spores (sporangia).

Rain (presence of free water) then becomes the principal factor in the development of the disease. During heavy rainfall, the spores are dispersed to the leaves and fruit by water splash where they release another type of spore (zoospores, motile spores) which infect grape tissue.

Young grape tissue tends to be more susceptible to downy mildew than older tissue. Once leaves are fully expanded they tend to be less susceptible to infection. Fruit are susceptible only from bloom through 4 weeks post-bloom. However, since new succulent leaves are produced throughout the growing season, it is important to maintain good fungicide coverage for downy mildew.

SCOUTING

Grape varieties: Monitor susceptible varieties first.

When: From first leaf unfolded (stages 5-7), particularly during periods of rain, heavy dew and persistent fog at temperature above 11°C.

Where: Give priority to sections of the vineyard that remain wet: heavy soil, poor drainage, depression, abundant foliage. Downy mildew tends to develop on the shadiest side of the canopy first.

How: Monitor for the appearance of oil spots on the upper surface of the leaves. A white, downy sporulation that can be removed with the fingers can be seen on the lower leaf surface. Check for infected shoot tips, which bend in a "shepherd's hook", and become white with sporulation.

Early in the season, it is sometimes difficult to differentiate between downy mildew and normal "blotchiness" of young leaves. Suspected downy mildew lesions can be confirmed by placing leaves with symptoms in a plastic back with damp paper towel over night. Downy white sporulation will be evident on the underside of oilspots if it is downy mildew.

POWDERY MILDEW

Fungus: Erysiphe necator; (Uncinula necator) (Blanc, Oidium)

SYMPTOMS

Leaves: The first powdery mildew lesions are frequently found on the undersides of leaves (a). As the epidemic progresses, lesions become apparent on the upper sides of leaves as well. These lesions will increase in size and number if the disease is left unchecked. Severely infected leaves may become brittle and drop off. Starting as early as late July, very small orange to black spherical structures called cleistothecia develop on the upper and lower surfaces of leaves (b).

Shoots: Brown to black irregular blotches that can measure up to a few centimetres, follow the gradual degeneration of the fungus over the course of the season (c). The spots have indistinct margins and remain visible even following shoot lignification.

Inflorescences and rachis: Usually seen on rachis, powdery mildew has the appearance of a grey to whitish powder. Severe infections of the rachis can result in clusters being dropped, especially if mechanical harvesting is done. Symptoms on the rachis are similar to those on shoots

Berries: Berries can be infected from immediately after bloom through 4 weeks post-bloom.

They turn an ash grey colour and quickly become covered in spores (d), giving them a floury appearance. At the end of the season, cleistothecia also appear on the berries (e). A ffected berries dry out and may drop off (f). Berries that infected later during the period of susceptibility are prone to splitting, making them susceptible to infection by *Botrytis*.

Powdery Mildew Symptoms

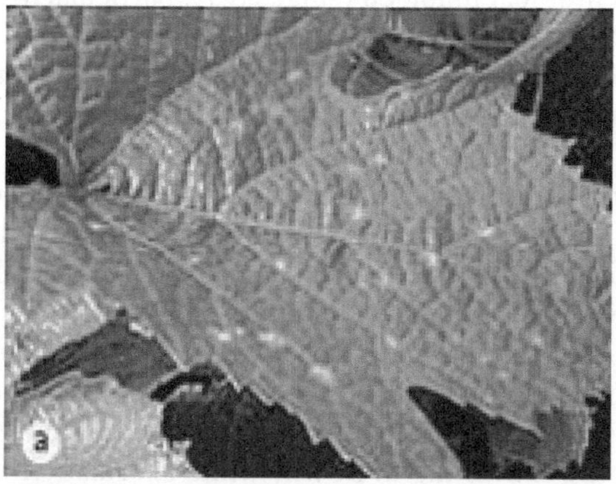

The first powdery mildew lesions are frequently found on the undersides of leaves

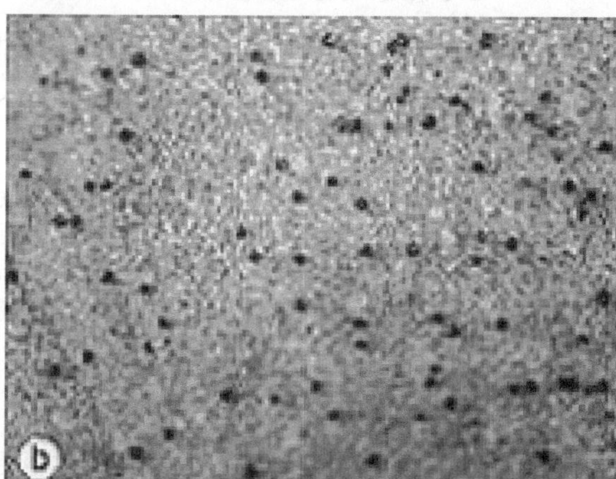

Very small orange to black spherical structures called cleistothecia develop on the upper and lower surfaces of leaves

The gradual degeneration of the fungus over the course of the season

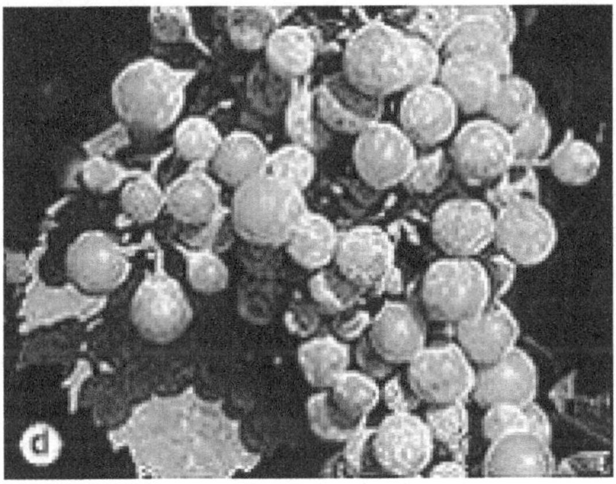

They turn an ash grey colour and quickly become covered in spores

Cleistothecia also appear on the berries

Affected berries dry out and may drop off

EPIDEMIOLOGY

E. necator is an obligate parasite of grapevines, i.e., it can develop only on living grapevine tissue. In our climate, *E. necator*overwinters as cleistothecia, structures containing ascospores (or sexual spores). In the spring, the ascospores mature and infect the leaves growing in proximity to bark. Following infection, spots covered with asexual spores, called conidia, develop on the leaves. The conidia of *E. necator* do not need free water on the tissue to infect it. However, high relative humidity promotes germination of the conidia and therefore infections. Powdery mildew of grape is promoted by hot (optimum temperature of 25°C), dry (but humid) weather since water inhibits germination of the conidia.

SCOUTING

Grape varieties: Monitor susceptible varieties.

When: Start looking for powdery at about 3-5 leaves and continue throughout the season. *Where:* Throughout the vineyard.

How: Monitor for the appearance of discolored spots that turn whitish on the upper and lower leaf surfaces. Do not confuse with pesticide residues. When checking, change the angle of the leaf. The whitish spots are particularly visible when the leaf is held at an approximately 30° angle. Be sure to sample leaves from the interior of the canopy since disease development is favoured by shade.

ANTHRACNOSE

Fungus: Elsinoe ampelina (Anthracnose)

SYMPTOMS

Leaves: The fungus will cause small round spots (a) which, as they age, give way to small holes (leaving a 'shot-hole' appearance)(b). During severe infections, the leaves shrivel up and drop.

Shoots: Deep elongated cankers, greyish in the centre with a black edge (c).

Inflorescences: Inflorescences are highly susceptible. During severe infections, they can turn yellow, brown, then dry out completely (d).

Berries: Deep spots, violet turning greyish in the centre, with a black edge (e). Severely infected berries dry up and drop prematurely (f, g).

Leaves: The fungus will cause small round spots

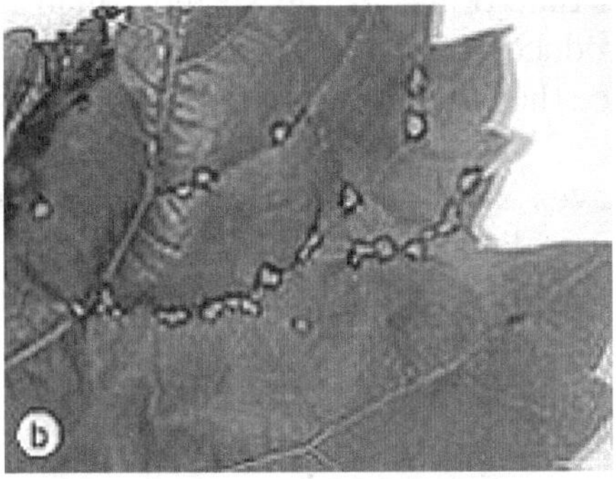

Leaves: As they age, they give way to small holes (leaving a 'shot-hole' appearance)

Shoots: Deep elongated cankers, greyish in the centre with a black edge

Inflorescences: During severe infections, they can turn yellow, brown, then dry out completely

EPIDEMIOLOGY

The causal organism of anthracnose, *Elsinoe ampelina*, probably overwinters in cankers formed on the canes and in infected berries on the ground and on berries left in the trellis. In the spring, small fruiting bodies, called acervuli, form and produce spores (conidia). The spores are covered with a mucilaginous substance that enables them to stick to the site of infection. They are dispersed by rain.

The longer the leaves and stems remain wet (>12 hours), the more severe the infection. The disease develops at temperatures of 10 to 35°C; the optimum temperatures for disease development are between 20 and 26°C. Symptoms appear 4 to 12 days after infection.

SCOUTING

Grape varieties: Susceptible varieties.

When: Particularly from stage 1st leaf unfolded (7) to stage 4-6 leaves unfolded and inflorescences visible (12). Monitor until the harvest, particularly after rainfall and storms, when the vines are wet for more than 12 hours.

Where: Entire vineyard.

How: Monitor for the presence of small black spots, particularly on young leaves.

GREY MOLD

Fungus: Botrytis cinerea (Botryotinia fuckeliana) (Pourriture grise)

SYMPTOMS

Botrytis cinerea is found in most fruit crops, including grapes.

Leaves: Although the most obvious symptoms are on the berries, *B. cinerea* can infect the green leaves and cause necrotic brown spots (a).

Inflorescences: Inflorescences can also be infected (b), causing the inflorescences to dry out or latent infections visible only at veraison.

Berries: Infected berries become covered with a greyish felt-like substance consisting of spores of the fungus (c, e). At that stage,*B. cinerea* is capable of directly infecting the berries(d, f), which become more susceptible as they mature and their sugar content increases.

B. cinerea can infect the green leaves and cause necrotic brown spots

Inflorescences can also be infected

Infected berries become covered with a greyish felt-like substance consisting of spores of the fungus

EPIDEMIOLOGY

The causal fungus, *Botrytis cinerea*, overwinters as sclerotia (compact mass of mycelium) or mycelium in plant debris. In spring, spores (conidia) form and cause lesions on young succulent grape tissues. Generally, *B. cinerea* requires free moisture to germinate and a wound to enter tissue. New spores are produced on the lesions. T

he spores are dispersed by the wind and cause several new cycles of infection, enabling the fungus to remain in the vineyard. Berries may be infected shortly after bloom but remain symptomless until veraison. Berries become more susceptible as they mature and their sugar content increases so*B. cinerea* is capable of directly infecting the berries without injury. When the fungus population is low, the damage generally does not result in economic losses. However, if the season is warm and wet (optimum temperatures of 15-25°C), and if grape berry moth is present, there is a greater chance of a severe infection, resulting in reduced yield and quality.

SCOUTING

Varieties: Particularly in susceptible varieties with very compact, thin-skinned clusters.

When: From onset of bloom (stages 19-21) and then during veraison (stages 35 and +).

Where: Entire vineyard.

How: Monitor drying out of the inflorescences and, during veraison, the presence of rotten berries with or without grey sporulation.

BLACK ROT

Fungus: Guignardia bidwelli (Pourriture noire)

SYMPTOMS

All aerial parts of the plants can be infected by this disease. Black rot may have a major impact on yield if the disease is not properly controlled.

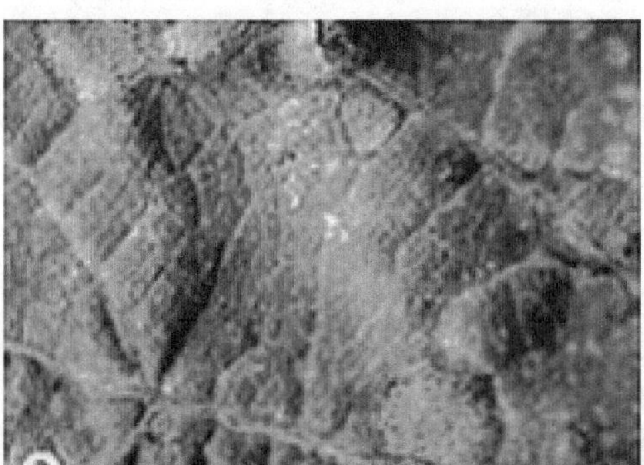

Leaves: presence of small brown lesions (2 to 10 mm in diameter) surrounded by a darker margin a ring of small black fruiting bodies (black pustules)

Shoots: under high disease pressure, brown to black elongated lesions with black pustules may appear.

Leaves: presence of small brown lesions (2 to 10 mm in diameter) surrounded by a darker margin, a ring of small black fruiting bodies (black pustules) (a).

Berries: At first, the berries become whitish then purple to black (b). Berries will dry and become mummified but most of the time will stay on the rachis. At the end of the season, berries will be covered by black pustules(c).

Berries: At first, the berries become whitish then purple to black

Berries: At the end of the season, berries will be covered by black pustules

EPIDEMIOLOGY

The fungus overwinters in mummified berries or tendrils within the vine or on the ground as fruiting body (perithecia) containing sexual spores. In the spring, these spores will be ejected from these fruiting bodies by rain and will infect new growth. The lesions will appear about 2-3 weeks after the infection (rain) on leaves and 10 to 14 days on berries. Black fruiting bodies containing spores (secondary spores) will develop on lesions and these spores will be disseminated to all new growth, leaves, petioles, shoots, tendrils and berries causing new infections until the end of the season. Infection will occur if the wetness last at least 6 hours and the temperature is between 9 and 32°C with an optimum at 20-25°C.

SCOUTING

Grape varieties: Monitor susceptible varieties especially *V. vinifera*. Refer to the Ontario Ministry of Agriculture publication *Fruit Production Recommendations* (publication 360).

When: Black rot generally appears after downy and powdery mildew, scouting should start at the 2-3 or 4-5 fully expanded leaves (stages 9-12). Scouting should be more intense during rainy periods combined with temperatures above 9°C. The critical period for berries infection is from end of flowering to beginning of veraison.

Where: Entire vineyard.

How: Monitor for the presence of small brown lesions on leaves and fruit rot.

CROWN GALL

Bacteria: Agrobacterium vitis (Tumeur du collet)

SYMPTOMS

Crown gall occurs in several wine producing regions. It is a serious disease because it is difficult to control and can cause the death of the vine.

Entire plant: The leaves and shoots wilt. The entire vine or only certain canes die (a).

The leaves and shoots wilt. The entire vine or only certain canes die

Base and canes: Galls, more or less spherical (b). Initially, the galls are whitish, spongy to firm turning dark and brittle. The size of the galls varies.

The surface of the galls is irregular and the epidermis of the stem peels (c, d, e). Galls are generally found at and just above the graft union but can develop on canes as well. Shoots and leaves may begin to wilt, especially during periods of high water demand by the vine.

Galls, more or less spherical

EPIDEMIOLOGY

The causal bacterium is often introduced into the vineyard via nursery stock. This disease is more severe in areas where winter injury is common. *A. vitis* can survive in the vascular system of the vines for several years without producing any symptoms. When cells are damaged by freeze injury or other factors, the bacterium invades plant cells associated with the plant's natural wound-healing process. Crown galls can take several years to appear. They develop normally at temperatures of 20 to 32°C and more slowly at temperatures of 15° and below. *A. vitis* is also capable of surviving for up to 5 years in root debris after infected vines have been removed.

SCOUTING

Grape varieties: All varieties, particularly *V. vinifera*.

When: At time of planting and early in the summer for vines older than one year. It is important to scout during summers that follow a very cold winter and/or a winter with little snow cover.

Where: The lowest areas of the vineyard, where frost pockets and poor soil drainage may occur.

How: Note the presence of wilting, particularly during periods of drought. A proliferation of sucker growth or aerial roots is also indicative of crown gall. Inspect the base of the plants for the presence of galls.

Select links below for corresponding photos

(1) Winter dormancy

(2) Bud swelling

(3) Wool (doeskin stage)

(5) Bud burst

(6) Green shoot
(7) First leaf unfolded
(9) 2 to 3 leaves unfolded
(12) 4 to 6 leaves unfolded, inflorescence clearly visible
(15) Inflorescence elongating; flowers closely pressed together
(17) Inflorescence fully developed; flowers separating
(19) Beginning of flowering; first caps falling
(21) Early flowering; 25% of caps fallen
(23) Full flowering; 50% of caps fallen
(25) Late flowering; 80% of caps fallen
(27) Fruit set; young fruits beginning to swell
(29) Berries small; bunches begin to hang (4-6 mm)
(31) Berries pea-sized; bunches hang (7-10 mm)
(33) Beginning of berry touch
(35) Beginning of berry ripening (veraison)
(38) Berries ripe for harvest

Winter dormancy

Wool (doeskin stage

Bud burst

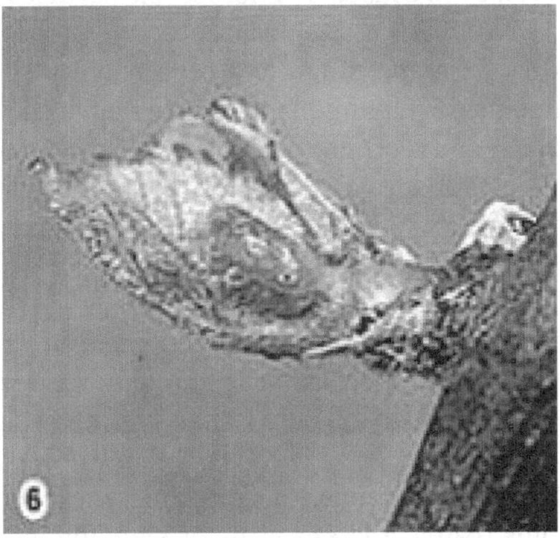

Green shoot

4

Grape Virus Diseases

WHAT IS A VIRUS?

Grape virus diseases are caused by microscopic particles that are composed of genetic material (DNA or RNA) inside a protective protein coat. Viruses are much smaller than bacteria and fungi, and require very specialized tests for their detection. Once inside a living plant, viruses have the ability to multiply by taking over plant cells and reprogramming them to make more virus particles. In grapevines the virus will spread systemically to all parts of the vine (roots and vegetative parts). Once a plant is infected, it will remain so for life.

GENERAL VIRUS SYMPTOMS

Viruses cause a wide range of symptoms, ranging from no visible symptoms to plant death. Many grapevine viruses cause a general decline in vigour and productivity and delayed maturity of fruit. Other symptoms may also be present on leaves, stems or fruit.

It is possible for a virus to infect a grapevine without the plant showing any obvious symptoms (called a latent virus or sleeping virus). The degree of virus symptom development and effect on a plant is influenced by the virus strain, plant variety and age, environment, and time of year. A previously latent virus may become more virulent (or more severe) at some time due to changes in the environment or through propagation, often onto a different rootstock. The symptoms may be subtle, requiring proper experiments to demonstrate the effects of the virus. For example, the involvement of a virus may not be obvious in an increased susceptibility to winter injury or a gradual decline in yield. Symptoms such as graft rejection, rapid decline of vines, severe stunting, late blossom and late or poorly ripening fruit are more obvious. Foliar, fruit or cane symptoms of virus diseases are also frequently visible. These may include any one of the following: red foliage (on red varieties), short internodes, mottling of leaves, fasciation of canes, double internodes, excessive growth from secondary buds, straggly bunches with both large and small berries, rolling of the leaves in the fall, wood pitting and grooving, etc.

IMPORTANCE OF VIRUS DISEASES

Virus diseases can have a serious impact on vine health, yield and quality of the fruit. Symptoms are not always severe or obvious, but even a small decrease in yield will add up over time causing significant economic losses. Decreased yields of 5 to 10% are not uncommon for grapevine viruses, and losses can be much higher. Viruses may also influence wine quality by causing delays in sugar accumulation, poor acid development, and poor colour development.

The difficulty of detecting virus infection can lead to rapid dissemination of virus-infected material through propagation. When infected cuttings are used for propagation, whole vineyards can become infected.

HOW VIRUSES ARE SPREAD

Humans are the most effective transmitter of virus diseases through movement and propagation of virus infected rootstocks, cuttings and finished plants.

Most viruses also have other methods of spreading, although not all are known. Insects or other organisms that spread viruses are known as "vectors". Many serious grapevine viruses are classified as "nepoviruses" which are transmitted by certain species of nematodes. Examples of nepoviruses include Grapevine fanleaf virus, Arabis mosaic virus, Tomato black ring virus, Tomato ringspot virus, Tobacco ringspot virus, Peach rosette mosaic virus and many others. These viruses can cause significant economic losses. The nematode feeds on the roots of infected vines and retains the virus for several months. The disease is spread as the infected nematode feeds from grapevine to grapevine. Infected nematodes may be spread on the roots of nursery plants and through soil water (irrigation, seepage, flooding). Fortunately plant protection measures that require treatment of imported planting stock have prevented the most damaging virus-vectoring nematode of grapevines, *Xiphinema index*, from being introduced into British Columbia. However, other nematodes found in British Columbia, such as *Xiphinema americanum*, are known to vector nepoviruses. Some of these nepoviruses may also be spread through seeds of infected weeds in the vineyard. They may then be picked up by the nematodes and moved from the weeds to grapevines.

Other viruses are transmitted by soil fungi or insects such as aphids, leafhoppers, psyllids and mealybugs. Mealybugs and soft scale insects are known to be vectors of some of the grapevine leafroll-associated viruses, *Grapevine virus A*(causing Kober stem grooving disease, formerly *Grapevine stem grooving virus*), and *Grapevine virus B* (causing grapevine corky bark disease).

There are also virus-like diseases for which the method of transmission (apart from propagative) is not known or well documented. This group includes many damaging diseases such as Grapevine Enation, Grapevine Yellow Speckle, and Grapevine Shoot Necrosis. Little is known about some of these viruses beyond the symptoms they cause. This lack of knowledge forms a serious obstacle to effective disease management.

CONTROL AND PREVENTION OF VIRUSES:

- Prevention of virus diseases is critical, as there are no cures for virus-infected vines other than vine removal.
- Plant only fully virus-tested vines to reduce the risk of introducing virus diseases into your vineyard.
- If an insect vector is known to spread a virus, controlling the insect may help to limit or reduce the rate of spread within a vineyard.
- Canada has specific grapevine import regulations that help to protect grape growers from the introduction and spread of virus diseases and other pests which are not yet established in British Columbia. Selected grapevine varieties/clones and rootstocks from Canadian-approved nurseries in France and Germany are currently approved for importation into Canada. Under this program tests for regulated viruses/pests of grapevines are carried out both in the exporting country and in Canada, prior to the approval for the importation of any variety and rootstock. Canada has different import requirements for grapevines from the U.S, because of the similar pest situation and the proximity of the two countries. For more information on importing grapevines, contact the Canadian Food Inspection Agency (CFIA).
- Note that important viruses such as *Arabis mosaic virus*, *Grapevine fanleaf virus* and the viruses causing grapevine leafroll disease have been removed from the federal quarantine pests list. Although CFIA-approved foreign certification programs are supposed to produce plants free of these viruses, their removal means that foreign nurseries approved to ship grapevines into Canada no longer have to prove their vines are free of these viruses. Grapevines infected with these viruses have been imported since this de-regulation. It is recommended that growers importing vines request that they be tested for these viruses in the country of origin. These viruses cause serious economic diseases.

CLONAL SELECTION

Elimination of viruses from grapevines is possible, but requires a combination of heat therapy and micro-shoot tip tissue culture. This process is both expensive and time consuming, as it must be verified that the virus was successfully removed. Such vines then become the foundation for propagation programs designed to produce "clean" vines for industry. Vines that are to be subjected to virus elimination should be chosen carefully, so that only the very best varieties or clones are selected. The CFIA Laboratory at Sidney, British Columbia (Centre for Plant Health) provides testing, virus indexing and therapy for viruses of grapevines, tree fruits and other crops. These tested mother plants may then be placed in a Generation 1 (Nuclear) collection (repository) at the Sidney Lab or become mother plants for growers or nurseries. Propagative material from these mother plants may be used to produce other plants for certification under the Canadian Export Certification Program for the eventual establishment of healthy vineyards.

MAJOR GRAPE VIRUS DISEASES

There are over 70 different viruses and viroids that infect grapevines distributed throughout the world. Several are known to exist in Canada already. With increasing international trade and travel it is expected that more of these viruses will arrive.

A national grapevine virus survey in 1994-1995 found vines infected with Arabis mosaic virus, Grapevine fanleaf virus and Grapevine leafroll-associated viruses 1 and 3 in both Ontario and British Columbia. Leafroll 3 was the most commonly found virus, but the other viruses were present at very low levels in British Columbia. For example, out of 1485 B.C. samples, only 1 positive was found for fanleaf virus, and only 5 positives for arabis mosaic. Although the survey was not detailed enough to determine incidence, the low number of positives is an encouraging indication that B.C. remains largely free of these damaging viruses, and it is worth-while to continue to prevent introduction through careful screening of nursery stock. However, additional grapevines infected with these viruses have been imported since 1995.

GRAPEVINE LEAFROLL-ASSOCIATED VIRUSES

Grapevine leafroll is probably the most wide-spread virus disease of grapevines world-wide. There are currently 5 different viruses associated with leafroll, including:

- *Grapevine leafroll-associated virus 1* (GLRaV-1),
- *Grapevine leafroll-associated virus 2* (GLRaV-2),
- *Grapevine leafroll-associated virus 3* (GLRaV-3),
- *Grapevine leafroll-associated virus 4* (GLRaV-4), which includes several different strains
- *Grapevine leafroll-associated virus 7* (GLRaV-7),

GLRaV-3 has been the most commonly found leafroll virus in Canada, however GLRaV-1 and GLRaV-2 are also present. In the national grapevine survey conducted in 1994-1995, GLRaV-1 was present in 1.0% of samples tested, whereas GLRaV-3 was present in 10.5% of samples tested nationally. Preliminary results from current surveys of symptomatic vines in conducted by Agriculture and Agri-Food Canada, Summerland, in 2011-2013 indicated that GLRaV-3 and GLRaV-2 are the predominant leafroll viruses in B.C. Smaller amounts of GLRaV-1 and GLRaV-4 were also found.

Symptoms: Leafroll virus has impacts on both vine health and grape quality. Growth and yield may be reduced by 10-40%. The virus reduces yield by inhibiting cluster formation and development. Infected vines have an increased sensitivity to environmental stress. Quality impacts included delayed maturity of grapes, a 25-50% reduction in sugar content, poorly coloured fruit, and modified aromatic profiles of the must.

Leafroll symptoms are highly variable between cultivars. Typical leaf symptoms include reddening of the leaves between major veins in red varieties, and yellowing of the leaves between major veins in white varieties. In both cases, a key diagnostic symptom is that primary veins usually remain green in colour. Leaves become thick, brittle, and roll downwards in some cultivars. Symptoms typically appear around veraison, beginning with older leaves and gradually moving upwards in the canopy as the season progresses.*Grapevine leafroll disease. Photos courtesy Canadian Food Inspection Agency.*

Transmission: Long distance spread occurs primarily by movement of propagation material from infected plants. Several mealybug and soft scale insect species are known to transmit leafroll associated viruses 1, 3 and 4. Leafroll vectors that are present in B.C. include the common grape mealybug (*Pseudococcus maritimus*), the cottony maple scale and the European fruit lecanium scale. There are no known vectors for GLRaV-2 or GLRaV-7.

Management:

- Source vines that have been tested free of leafroll viruses, if available.
- Monitor for and manage leafroll vectors including mealybugs and soft scales.
- Monitor for leafroll symptoms in late summer. Since symptoms are often variable and unreliable, have suspect vines tested to confirm the presence of leafroll.
- If a small percentage of vines in a block are known or suspected to be infected, it is best to remove them to limit spread within the block.
- If a high percentage of vines are infected, consider whole block removal.
- Take measures to prevent moving mealybug and scale insects from infected to healthy blocks by working infected blocks last and cleaning equipment between blocks.

Grapevine Fanleaf Virus

Fanleaf degeneration disease is thought to be the most serious grapevine virus disease. The severity of symptoms varies by cultivar.

Symptoms: Infected leaves are often malformed with abnormally gathered primary veins, giving the leaf the appearance of an open fan. Other symptoms may include a yellow mosaic pattern on leaves or bright yellow bands along major veins. Fan-shaped leaves may or may not be present with mosaic or veinbanding symptoms.

Grapevine fanleaf virus symptoms - yellow mosaic pattern on leaf. Photo courtesy Canadian Food Inspection Agency. Grapevine fanleaf virus symptoms - bright yellow vein banding on leaf. Photo courtesy Canadian Food Inspection Agency.

Affected vines tend to be smaller than healthy vines. Sensitive varieties show progressive decline, low yields (up to 80% losses) and low fruit quality. The productive life of the vineyard is shortened and winter hardiness is decreased.

Grapevines with yellow leaf mottle due to Arabis mosaic nepovirus, cv. Tamiioasca romineasca.Photo courtesy A. Eppler, Justus-Liebig Universität, www.ipmimages.org

Transmission: Fanleaf virus is transmitted by the nematodes *Xiphinema index* and *X. italiae*, which have never been found in Canada. There are no natural weed hosts. Long distance spread occurs primarily by movement of propagation material from infected plants.

Arabis Mosaic Virus

Symptoms: The most common symptoms include leaf mottling and flecking, stunting and leaf deformation including enations. Many infections are latent and plants do not show symptoms. Symptoms can vary greatly by variety, rootstock and environment conditions. Yield losses of up to 50% may occur through reduced growth, dieback and severe dropping of fruit.

Transmission: Long distance spread occurs primarily by movement of propagation material from infected plants. Arabis Mosaic Virus is transmitted by the nematode *Xiphinema diversicaudatum*, which has been found in scattered locations in Canada. This virus can also infect many other herbaceous and woody hosts such as raspberry, strawberry, rhubarb, cherry, peach, and plums.

RUGOSE WOOD COMPLEX VIRUSES

The rugose wood complex is caused by several viruses, including *Rupestris stem pitting associated virus* (RSPaV),*Grapevine virus A* (GVA) and *Grapevine virus B* (GVB).

RSPaV is a common and widespread virus of grapevines. It is not regulated in Canada or the USA. Both GVA and GVB are regulated diseases in Canada, and have been found in only a few localized areas. GVB has not been found in B.C. to date (2014).

Symptoms: GVA is associated with Kober stem grooving disease. Affected vines may show swelling at the graft union and fail to thrive. GVB is associated with corky bark disease, which may cause an incompatibility to develop at the graft union. Leaf symptoms resemble those of leafroll virus but are more severe, and severe stunting of the vine may occur.

Rupestris stem pitting causes a slow decline in growth, resulting in vines that are smaller than normal with reduced yields. No leaf discoloration is observed. The disease affects only grafted vines; ungrafted vines may be infected, but usually do not show symptoms. Small pits may develop in the wood on rootstocks, in particular *V. rupestris* and American rootstocks. The severity of the disease is more pronounced in vines that are also infected with other viruses of the rugose wood complex, including GVA and GVB.

Transmission: Spread is mainly through propagation. There is no known insect vector for Rupestris stem pitting virus. GVA and GVB can be transmitted by some species of mealybugs. The cottony maple scale (*Neopulvinaria innumerabilis)*can transmit GVA.

Management: Disease management depends upon use of certified healthy, virus-free stock. Remove vines showing suspicious symptoms.

GRAPE DISEASES

BOTRYTIS BUNCH ROT (*BOTRYTIS CINEREA*)

Botrytis bunch rot, caused by the fungus *Botrytis cinerea*, is a common problem wherever grapes are grown. The disease can cause serious losses in both yield and quality when weather conditions favour the disease.

Symptoms:

Discoloured, shrivelled berries caused by Botrytis Bunch Rot. Photo courtesy P.Sholberg, Agriculture & AgriFood Canada.

Bunch rot of grape cluster caused by Botrytis cinerea. Note light grey sporulation of the fungus on infected berries.

Ripening grapes are affected by a rot which may progress to infect whole clusters. With sufficient rain and humidity, berries split open and develop a greyish mold on the surface. Affected berries may shrivel in the dry Okanagan climate.

Botrytis may cause girdling lesions on the pedicel or rachis, leading to drying of clusters or portions of clusters, and premature cluster drop.

Infection and Spread:

Botrytis infects grape shoots, flowers, leaves and fruit under a range of temperature and humidity conditions (Table 1). Infection is optimal at 15-20 °C with free water or over 90% humidity. Grape cultivars with dense canopies, thin skins, and/or tight clusters are more susceptible to botrytis bunch rot (Table 2).

Botrytis overwinters as mycelium or sclerotia on bark and as mycelium in dormant buds. It may also overwinter in mummified fruit. In the spring, conidia (spores) are produced which can infect leaves and young clusters before bloom. The fungus may infect blossoms leading to fruit infection, but the fungus becomes inactive (latent) in the fruit possibly due to low sugar and high acid contents. It becomes active again when the berries begin to soften. Fruit infection usually begins in berries with 5 to 8 percent sugar (at veraison), with berries remaining susceptible up to harvest and also post-harvest. Fruit become increasingly susceptible has they mature.

Abundant conidia are produced on infected fruit, leading to secondary spread to adjacent berries in a cluster and to nearby clusters. Spread of the spores is aided by summer rains, overhead irrigation, heavy dew and juice from split berries. Spores are also carried by the wind. Mechanical damage caused by birds or insects creates entry points for the fungus and can greatly increase losses due to bunch rot.

Table. ***Time/temperature relationships for botrytis bunch rot infection***

Temperature	Amount of Time
10 °C	30 hours or more
15.5 °C	18 hours or more
22.5 °C	15 hours or more
26.5 °C	22 hours or more
30 °C	35 hours or more

Slow drying conditions together with high humidity (90% or better) favour the development of *Botrytis*. Table 1 indicates the temperature required plus the corresponding period that slow drying conditions must exist for bunch rot infection to occur.

Cultural Control of Botrytis Bunch Rot:

During the growing season:

- Avoid overhead irrigation and keep irrigation periods as short as possible.

- Prevent excessive vine growth by judicious use of water and fertilizer.
- Canopy management, including shoot thinning and leaf removal before fruit set will reduce botrytis bunch rot. Remove leaves and lateral shoots located opposite, one node above, and one node below each fruit cluster.
- Prevent berry damage by effective control of powdery mildew early in the growing season.
- Minimize berry damage by birds and insects.

During the dormant season:

- Reduce carry-over of botrytis to the next season by removing any remaining fruit from the vines during pruning. If practical, remove trash from the vineyard floor.

Table . Susceptibility of grape varieties in British Columbia to botrytis bunch rot

Very Susceptible	Susceptible	Moderately Susceptible	Least Susceptible
Auxerrois	Barbera	Ortega	Cabernet Franc
Bacchus	Kerner	Scheurebe	Cabernet Sauvignon
Chardonnay	Lemberger	Schoenburger	Chancellor
Chasselas	Malbec	Syrah	Foch
Chenin Blanc	Muscat Ottonel	Vidal	Merlot
Ehrenfelser	Pearl of Csaba	Viognier	Petit Verdot
Gamay Noir	Pinot Gris		
Madeleine Angevine	Pinot Meunier		
Madeleine Sylvaner	Pinot Noir		
Optima	Sangiovese		
Pinot Blanc	Sauvignon Blanc		
Riesling			
Semillon			
Siegerebe			
Zinfandel			

Chemical Control of Botrytis Bunch Rot:

Several fungicides are registered for the control or suppression of botrytis bunch rot. Alternate fungicides from different chemical groups to help prevent the development of resistance.

[1]Chemical Group: Products with the same number belong to the same class of compounds. Alternate products with different chemical groups to help delay or prevent the development of resistance.

[2]Pre-harvest Interval - time interval between last application and harvest

The number of spray applications necessary to control bunch rot depends upon disease pressure in the vineyard and weather conditions, as well as other factors such as susceptibility. Fewer applications may be needed if weather is very dry and/or disease pressure is low. Use a full program in vineyards where bunch rot was a serious problem in the previous year, and where sanitation measures were not taken after harvest. The number of sprays can be reduced in vineyards with a history of low disease levels and for all vineyards in dry years.

Table. ***Fungicides registered for bunch rot control on grape***

Fungicide	Chemical Group[1]	Active ingredients	Rate/ha	Rate/acre	PHI[2](days)	Comments
Rovral	2	iprodione 50%	1.5 kg/ha	600 g/acre	7	Maximum 2 sprays/season
Elevate	17	fenhexamid 50%	1.12 kg/ha	450 g/acre	7	Maximum 3 sprays/season
Pristine	7+11	pyraclostrobin 12.8% + boscalid 25.2%	420 – 735 g/ha	170-300 g/acre	14	Registered as suppression only for botrytis bunch rot. Maximum 2 sprays/season. Wine grapes only.
Luna Tranquility	7+9	fluopyram 125 g/L +pyrimethanil 375 g/L	1.2 L/ha	485 mL/acre	7	Maximum 2 sprays per season. Wine grapes only.
Scala	9	pyrimethanil 400g/L	2.0 L/ha	810 mL/acre	7	Maximum 3 sprays/season
Vangard	9	cyprodinil 75%	750 g/ha	300 g/acre	7	Maximum 2 sprays/season
Switch	9+12	cyprodinil 37.5% +fludioxonil 25%	775-975 g/ha	314-395 g/acre	7	Maximum 2 spays/season
Serenade MAX	NC	*Bacillus subtilis*	3-6 kg/ha	1.2-2.4 kg/acre	0	Biofungicide. Disease suppression only.

Key timings for *Botrytis* fungicide applications are bloom, just before bunch closing, veraison and pre-harvest. Botrytis control becomes more difficult as the grapevine matures because heavy canopy growth and bunch closing make it difficult to place the fungicide where it is needed.

Under typical Okanagan weather patterns, high risk periods for *Botrytis* infection are during bloom and from 1 to 2 weeks after veraison to harvest. These growth stages often coincide with rainy periods during June and September.

Be cautious about adding spreaders or stickers to fungicide sprays, as some spray adjuvants have been shown to increase the development of bunch rot on grape clusters. It is believed that spray adjuvants remove the natural wax protection on grape berries.

5

Grape Insect Pests and Diseases

DISEASES

Seriously, it does not matter if you have a small backyard project or a larger scale vineyard you will be faced with these less than pleasant elements of nature that you will need to know how to deal with.

But don't worry, it's very easy if you put the following disease and pest maintenance plan into action.

Even more, you should consult local agricultural departments and fellow grape growers in all aspects of canopy management, disease and pest control.

Regions vary greatly in disease and pest issues. A major problem in one area may not exist in another. Local knowledge will help you tremendously.

The two most effective means of disease control are good canopy management and preventative fungicides treatments.

You will need to use fungicide on a regular basis and it should be done in a preventative maintenance manner. Waiting until fungus appears is usually too late. Be sure to visit the educational sites I reference later in the fungus and insect sections. They are chalk full of healthful information. This information is an easy read short and to the point with excellent images. Spend a few minutes to an hour on these sites and you'll practically be an expert.

Canopy management insures a uniform vine structure. The canopy includes all the stems, leaves and fruit clusters that have grown from your two main canes. Basically canopy management is the thinning and positioning of leaves, stems and fruit clusters during the summer. Positioning includes spacing shoots and fruit clusters amongst your catch wires so that all have equal exposure to sunlight and air flow.

Removing fruit clusters is an emotional experience but can help produce significantly higher wine quality.

Wine that has been produced from grapes at different stages of ripeness will be inferior to that produced from well managed canopy systems. The more sunshine and air flow that is allowed to penetrate your vines the less likely you will encounter fungus diseases.

Thinning:

- remove shoots that are growing downward
- remove shoots from excessive areas of vine density, a general rule of thumb is to have two to four shoots per foot of trellis
- remove shoots that are not bearing fruit, unless they are being retained for possible cane renewal
- remove shoots that have emerged from the trunk base, unless they are being retained for possible cane renewal
- remove excess leaf growth in fruiting zones so that grapes can dry easily and have good sun exposure

Each shoot should contain no more than one grape cluster at an average size of 5-8 ounces. An average size cluster needs approximately ten to fifteen leaves for proper ripening. If your grape clusters are small you can get by with two per shoot. For uniform results all shoots should have approximately the same amount of leaves.

Positioning:

In the growing season use twine to insure a uniform canopy by securing the shoots evenly and upwardly to the catch wires. Again, position all shoots and fruit clusters for equal exposure to sunlight and air flow. Be sure to position fruit clusters so that they are not touching one another.

FUNGUS:

Powdery Mildew: Grape vines infected with powdery mildew display white powder-like splotches on leaves, stems and grapes. Powdery mildew is experienced worldwide. It can grow well in both wet and dry regions. It can kill leaves and defoliate the vine. Grape quality suffers when leaves are unable to perform proper photosynthesis. A well managed canopy and regular sulfur spraying will prevent most powdery mildew occurrences. Spray every ten days when shoots reach 10" in length, continue up until thirty days before harvest: More on Powdery Mildew, Cornell University

Downy Mildew: Symptoms include light green to yellow spots scattered across the leaf. These spots appear greasy and are commonly referred to as oil spots. The biggest concern for downy mildew is leaf infection. The fungicides Copper, Captan, and Mancozeb can be sprayed every ten days and is generally mixed with sulfur. Please read the labeling and do your research, local restrictions may apply:More on Downy Mildew, Iowa State University

Black Rot: Symptoms include brown circular lesions on infected leaves. Left untreated it can destroy an entire grape crop. The biggest concern for black rot is the infection of young grape clusters.

Infected berries first appear light brown and then turn to near black as masses of black pycnidia develop on the surface. Again, Copper, Captan, and Mancozeb are the most protective fungicides. For application follow the downy mildew recommendations. Please read the labeling and do your research, local restrictions may apply: More on Black Rot, Cornell University

Bunch Rot: Infected berries appear soft and watery. In high humidity regions berries become covered in a grayish growth of fungus mycelium. Tight-clustered grape varieties are most vulnerable to bunch rot.

Canopy management is your most effective tool in preventing bunch rot. Promote good air circulation, sunlight penetration, and uniform leaf development by practicing proper canopy management. Remove leaves around grape clusters so that sunlight and air flow can help dry wet grape clusters: More on Bunch Rot, Ohio State University

INSECTS:

It's much easier to correct and control damage caused by insects then that caused by fungus disease. You can evaluate the problem as it develops and determine if damage is significant enough to warrant spraying. Rarely will insect populations be high enough to require insecticides.

Insects feed on buds, leaves, and fruit. The most damaging are those that feed on fruit because resulting rot can spread throughout the entire cluster.

Use insect specific insecticides to eradicate insects you have identified as damaging to your vines. Certain insects are actually good for your vineyard and you don't want to destroy them because they feed on mites and other problem insects.

Get local agricultural advice when seeking solutions for insect infestations. There's tons of good images and information regarding insect eradication online. Search by your state for local information.

Some of the most common and damaging insects are discussed and pictured at the following education sites:

More on Grape Insect Pest, University Of Minnesota

More on Grape Insect Pest, University Of Kentucky

Every state monitors insect populations most of which can easily be found online.

OTHER PEST:

In some areas birds can be a big problem. Their pecking of grape berries can easily initiate rot damage. Scare devices are helpful but can be annoying to neighbors if they are sound related. In some extreme cases netting may be your only solution. Voracious birds can devour a crop in a short period of time.

Deer, bear, kangaroos, raccoons just about every animal loves grapes. Deer the largest animal problem for most grape growers can be fenced out. Fence designs vary depending on what animal you are having problems with. Repellents such as hot sauce and pepper are effective in areas where problems are moderate. Again local knowledge is your best tool.

GRAPE INSECT PESTS OF THE HOME GARDEN

Growing grapes is becoming increasingly popular for the home gardener. Once grapevines are established, they can be productive for over 40 years. The productivity of the grapevines relies heavily on the careful choices of the home gardener. Home fruit gardeners can be successful if they select the right cultivars, properly prune grapevines annually, and maintain good fertility and pest management programs.

Accurate insect identification is essential in any pest management program. In this publication we have compiled a list of insects commonly found on grapes in Minnesota. Please remember that careful, timely inspection of grapes and good cultural practices will aid in reducing the need for insecticide applications.

FOLIAGE PESTS

Grape Phylloxera

Fig. Grape phylloxera.

Fig. Grape Phylloxera 2

Grape phylloxera, *Daktulosphaira vitifoliae* (Hemiptera: Phylloxeridae) possess both a root and foliar form. Although the two forms behave differently, both belong to the same species of phylloxera that occurs on the leaves and roots of grapes. In Minnesota, it is only the foliar form that potentially can cause damage as the rootstocks used in Minnesota are tolerant of feeding by the root form of phylloxera.

Identification and life cycle

Adult grape phylloxera are tiny aphid-like insects with a yellow body. The foliar form of grape phylloxera causes the formation of tiny galls, i.e. abnormal growths, to form on the leaves. Grape phylloxera survive the winter as eggs under the bark of the grapevine. In the spring, eggs hatch and nymphs move onto new leaves and develop new galls. Once mature, the female begins to lay eggs within the gall. Nymphs hatching from these eggs crawl to new leaves at shoot tips, settle on the leaves, and form new galls. There are several generations of phylloxera each season.

Damage

In most cases, grape phylloxera only affects the appearance of the leaves without impacting the health of the grape. Severe infestations of foliar phylloxera can reduce the photosynthetic ability of grape leaves. In addition, high levels of leaf galls cause leaf distortion, necrosis, and premature defoliation. Premature defoliation may delay ripening, reduce crop quality, and predispose vines to winter injury.

Management

Routinely monitor grapevines and be aware of any gall formations on grape leaves. The galls may not be visually appealing; however, grapevines can support small to moderate populations of foliar phylloxera without it resulting in decreased vine vigor.

Insecticidal

In general, management of grape phylloxera is not required. If grape vines have a history of severe grape phylloxera infestations, an insecticide application may be needed, although management is challenging. Apply during the bloom stage with an appropriate insecticide, such as carbaryl.

JAPANESE BEETLE

Fig. Japanese beetle.

Japanese beetle, *Popillia japonica* (Coleoptera: Scarabaeidae), feeds on nearly 300 species of plants, including fruits, vegetables, ornamentals, field and forage crops and various weeds. Although it is considered established in Minnesota, it is not necessarily common throughout Minnesota yet.

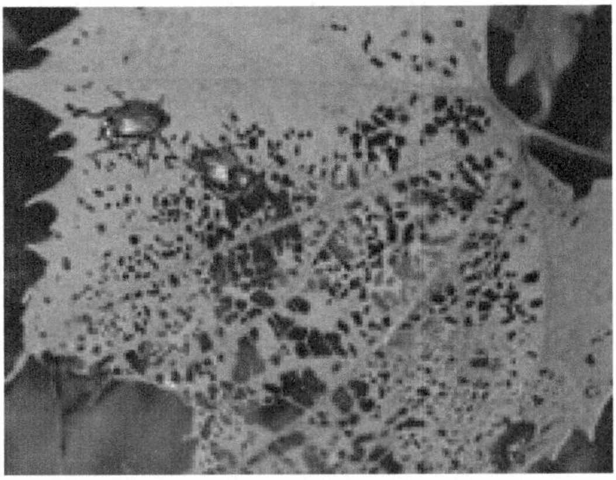

Fig. Japanese beetle damage on grapes.

Identification and life cycle

Adult beetles can be identified by their brilliant metallic green head and thorax with copper-brown wing covers and five patches of white hairs on each side of the abdomen and two patches of white hair on the tip of the abdomen. Eggs are very small (approximately 0.03 inch in diameter) and are laid in grassy areas. Larvae are typical white grubs, with a conspicuous brown head, legs, and a body bent into a c-shape.

Japanese beetle larvae overwinter as well-developed larvae in the soil. In the spring, larvae move toward the surface where they continue to feed on the roots of grasses and weeds and pupate.

Japanese beetle adults emerge from the ground in June and July and begin feeding upon foliage. Eggs are laid in the soil in July and August in grassy areas. Eggs begin to hatch in late July and larvae feed and grow until cold weather, at which time they tunnel 3 to 12 inches down and make overwintering cells. There is one generation per year.

Damage

Japanese beetle adults are leaf skeletonizers, i.e. they eat the leaf tissue between the veins, giving the leaf a lace-like appearance. Leaves fed upon by Japanese beetles soon wither and die. Grapes can tolerate small to moderate amounts of damage, while severely defoliated grapes can be injured.

MANAGEMENT

Examine plants at least twice per week during the growing season to check for Japanese beetle adults. Look for both adults and symptoms of their feeding.

Cultural

Japanese beetles prefer foliage exposed to direct sunlight, and vines with thin, smooth leaves, such as French hybrids, are preferred over those with thick, pubescent leaves. Therefore variety selection is important if Japanese beetle is of a concern to the home gardener.

Physical

If a few or a moderate number of adult beetles are found on grape foliage, handpicking them can provide acceptable control. The best time to physically remove Japanese beetles is in the morning before they become very active. Just knock them off into a pail of soapy water.

Insecticidal

If the population of beetles is such that handpicking them is not a feasible option, foliar applications of an insecticide such as permethrin, carbaryl, or malathion can help reduce the Japanese beetle population. There is no published threshold for Japanese beetle, however, if >15% of leaves are damaged; an insecticide spray may be warranted. It is important to note that Japanese beetles are gregarious and they may be present in great numbers on just a few vines. Therefore a localized spot treatment may be appropriate.

FLOWER AND FRUIT PESTS

GRAPE FLEA BEETLE

Identification and life cycle

Fig. Grape flea beetle and damaged grape bud.

Grape flea beetles, *Altica* spp. (Coleoptera: Chrysomelidae), are approximately 1/10 inch, long, shiny blue purple to blue green, with enlarged hind legs for jumping. Grape flea beetles overwinter as adults under the soil surface or in wood crevices, in and around the grapevines. In the early spring, adults emerge from overwintering sites, feed upon grape buds, mate and lay eggs. Larvae emerge in about two weeks and feed on leaves. Between late June and late July larvae will fall to the ground to pupate. Adult beetles emerge in late July and feed on grapevine leaves, but do not mate or lay eggs. In the fall the adults move to overwintering sites. There is one generation per year.

Fig. Grape flea beetle larva and damage on grape leaf.

Damage

Adult beetles damage primary buds when they feed on them, which prevents them from developing into shoots, which results in decreased grape yield. Larvae feed on the foliage but this is typically limited to several leaves and vines. Fortunately, larval damage does not usually affect grape quality or yield.

Management

Monitor grapevines in the early spring for grape flea beetle activity. Although grape flea beetles are active later in the summer, it is not important to monitor their activity then.

Cultural/Physical

Remove debris and leaf litter in and around grapevines. This will help to eliminate overwintering sites.

Insecticidal

In particularly heavy infestations, if >4% of grape buds are infested, spray an application of carbaryl, esfenvalerate, or pyrethrin to reduce grape flea beetles. Proper timing of the insecticide is imperative to provide good control of grape flea beetles.

GRAPE BERRY MOTH

Identification and life cycle

Fig. Grape berry moth.

Fig. Grape berry moth damage (note shriveled berries).

The adult grape berry moth, *Paralobesia viteana* (Lepidoptera: Tortricidae), is a small, inconspicuous, mottled brown-colored moth with a bluish-gray band on the inner halves of the front wings. It is approximately 3/8 inch long, with a wingspan of 1/3 to 1/2 inch. The newly hatched larva is creamy white with a dark brown head while late instars are green to purple in color, and are 1/3 inch in length when fully grown.

Adult moths emerge in mid to late May, mate, and lay eggs on or near grape flower clusters. Larvae hatch from eggs in 4-8 days and emergence of the overwintering generation peaks in mid-June and continues to mid-July. Larvae that hatch in June make up the first generation. Larvae feed on tender stems, blossom buds and berries. Some larvae pupate in the fruit cluster where they have fed. First generation adults begin to fly in late July, and the flight peaks in early August, however, adult moths continue to emerge until early September. Second generation larvae usually enter where the berries touch each other or where the berry joins the stem, and tunnel directly into the green berries and feed internally. Conspicuous red spots develop on the berries at the point of larval entry, and are referred to as "stung" berries. Larvae of the second generation complete their development in late September and pupate in fallen leaves. There are 2 generations/year.

Damage

The larva is the damaging stage of the insect, and causes damage by feeding on the grape berry. In addition, larvae can be a contaminant or act as an entry point for diseases. Injured berries ripen prematurely, split open and shrivel. Webbing produced by larvae prevents the berries from dropping. When grape clusters are ready to be picked, severely infested clusters may contain several larvae.

MANAGEMENT

Cultural/Physical

Because grape berry moth over-winters as pupae in leaf litter on the ground, removing and destroying dead leaves can help reduce grape berry moth emergence in the spring. In addition, covering leaf litter with one inch of compacted soil will prevent emergence. Both of these options must be completed three weeks prior to bloom. In light infestations, remove injured berries by hand.

Insecticidal

Where grape berry moth is an annual problem, a post bloom insecticide application may be necessary, and a late summer application may be needed to control the second generation. The number of spray applications depends on the amount of infested berries a home gardener is willing to accept. Effective insecticides include spinosad, pyrethrin, or carbaryl.

MULTICOLORED ASIAN LADY BEETLE

Identification and life cycle

The multicolored Asian lady beetle, *Harmonia axyridis* (Coleoptera: Coccinellidae), is a predatory lady beetle native to eastern Asia that was first detected in Minnesota in 1994. Multicolored Asian lady beetle adults are approximately 1/4 - 1/3 inch long and round. The coloration of multicolored Asian lady beetle adults is commonly orange to red with zero to 19 black spots, and rarely black with red spots. The most distinguishing feature of adult multicolored Asian lady beetles is the black "M"-shaped marking on the center of the pronotum (shield-shaped area behind the head). Eggs are yellow and oval-shaped. Multicolored Asian lady beetle larvae are alligator-shaped with black and orange markings, and are spiny in appearance.

Fig. Multicolored Asian lady beetle – adult (top left), larva (top right) and pupa (bottom).

Fig. Multicolored Asian lady beetle on damaged grape.

In the spring, multicolored Asian lady beetle adults leave their overwintering sites and mate. Adults then seek out colonies of aphids and lay eggs.

Larvae develop through four instars (stages). In the fall, multicolored Asian lady beetle adults move to grapes when their primary food source (aphids) is not available. At this time, grapes are close to maturity and may be injured due to splitting, or by wasps and birds, which creates feeding sites for adult beetles. Adult multicolored Asian lady beetles migrate from fields and wooded areas to buildings in the fall where they overwinter in cracks or spaces. There are 2 generations/year.

DAMAGE

In Minnesota, multicolored Asian lady beetle adults seem to only be attracted to grapes that have been previously damaged by wasps, birds, or physiological splitting. Once grapes are damaged, adult beetles feed on the injured areas of the berries. In addition, adult beetles are a contaminant in wine production. Multicolored Asian lady beetle adults are difficult to remove from clusters of grapes after they are picked; therefore some may be crushed with the grapes during processing. The flavor of the resulting wine is often tainted by the alkaloids contained in multicolored Asian lady beetles.

MANAGEMENT

Cultural/Physical

For home gardeners, the most important means of controlling multicolored Asian lady beetles is to remove injured grapes or clusters from grapevines. This will help to reduce the attractiveness and infestation of grape clusters by multicolored Asian lady beetles. Grapes should be picked late in the day, when beetles are less likely to be active. In addition, shake the clusters to dislodge the multicolored Asian lady beetles prior to picking.

Insecticidal

Insecticides labeled for home gardeners are not a very effective management option because they need to be applied relatively close to when the grape clusters will be picked, which is when the multicolored Asian lady beetles are present. An insecticide with a 1-day or 0-day pre-harvest interval, such as pyrethrin, may provide some control of multicolored Asian lady beetles. If multicolored Asian lady beetles are present when picking the grape clusters, it is imperative that they be removed prior crushing to prevent tainted wine or juice.

YELLOWJACKETS

IDENTIFICATION AND LIFE CYCLE

Yellowjackets, *Vespula* spp., (Hymenoptera: Vespidae) are approximately ½ inch long and have alternating yellow and black bands on the abdomen and are relatively smooth-bodied. Yellowjackets can be mistaken for honey bees because of their similar appearance and ability to sting. Honey bees are slightly larger than yellowjackets, are brown and black, and are covered with setae (hair). Foraging honey bees can also be identified by the pollen baskets on the rear legs that are often loaded with a ball of yellow or green pollen. The yellowjacket has a smooth stinger that can be used to sting multiple times, whereas the honey bee has a barbed stinger than can be used to sting only once.

Fig. Yellowjacket on damaged grape.

Fig. Yellowjacket on grapes.

Yellowjacket nests are constructed of several layers of combs made of tiny bits of wood fiber chewed into paper-like pulp. The colony rapidly increases in size and the number of adult yellow jackets may reach several hundred to several thousand by August.

DAMAGE

In late summer and early fall when their populations peak, the standard insect diet of the yellowjacket changes to sugar sources such as berries and grapes. Their feeding can cause damage to the grapes and can also be a danger because of their stings to people picking the grapes.

MANAGEMENT

Cultural/Physical

Grape clusters should be picked as soon as they ripen to discourage yellowjacket feeding. Remove any overripe or damaged fruit from the grapevines. In addition, do not leave any food items near the grapevines, as they may attract yellowjackets to the area.

Insecticidal

Insecticides are not an effective management option for controlling yellowjackets.

Table. Common names of insecticides and the pests they control

Insecticide	Example of Brand Name	Pest	Residual	Notes
Pyrethrin	Bug Buster-O	Grape flea beetle, grape phylloxera, multicolored Asian ladybeetle, grape berry moth	short	contact
Malathion	Spectracide malathion	Japanese beetle, grape phylloxera	short	contact
Spinosad	Monterey Garden & Insect Spray	grape berry moth	short	contact
Carbaryl	Sevin	Grape flea beetle, Japanese beetle, grape phylloxera	medium	contact
Permethrin	Eight Garden & Home Insect Control	Japanese beetle	medium-long	contact
Esfenvalerate	Ortho Bug-Be-Gon MAX	Grape flea beetle	long	contact

CAUTION: Always use insecticides strictly in accordance with label precautionary statements and directions. The label should state that the insecticide may be used on grapes or generally on fruit.Carefully observe the waiting period. Protect pollinators—do not spray grapes or other fruits when they are blossoming. If suggestions in this publication contradict label recommendations, the label is the final authority on how to use that specific product.

BLACK ROT (GRAPE DISEASE)

Grape black rot is a fungal disease caused by an ascomycetous fungus, *Guignardia bidwellii*, that attacks grape vines during hot and humid weather.

"Grape black rot originated in eastern North America, but now occurs in portions of Europe, South America, and Asia. It can cause complete crop loss in warm, humid climates, but is virtually unknown in regions with arid summers." The name comes from the black fringe that borders growing brown patches on the leaves. The disease also attacks other parts of the plant,"all green parts of the vine: the shoots, leaf and fruit stems, tendrils, and fruit. The most damaging effect is to the fruit".

Grape black rot affects many grape growers throughout the United States, therefore, it is important to understand the disease life cycle and environmental conditions to best manage the disease. Once infection takes place, different methods are available to control the disease.

DISEASE CYCLE

Guignardia bidwellii Disease cycle picture] The grape black rot pathogen overwinters in many parts of the grape vine and is able to over winter on the ground [especially in mummies]. In addition, pathogen can overwinter for at "least 2 years within lesions of infected shoots that are retained as canes or spurs."

Rains release the overwintering spores (Ascospores) that form within mummies on the ground and can be blown by the wind. Some of the mummies on the ground can have a significant discharge of ascospores that begins about 2 to 3 weeks after bud breaks and will mature 1-2weeks after the start of bloom. A second type of spore (Conidia)can also form within cane lesions or mummies that have remained within the "trellis, and these are dispersed short distances (inches to feet) by splashing rain drops."

Infection occurs when either of the spore types land on green grape tissue and tissue remains wet for a "sufficient length of time, which is dependent on temperature." The period that these overwintering spores are allowed to cause infection depends on the source.If there is a large source for infection, infection will set in early. In the presence of moisture, these ascospores slowly germinate, taking 36 to 48 hours, but eventually penetrates the young leaves and fruit stems (pedicels).

The infections become visible after 8 to 25 days. When the weather is moist, ascospores will be produced and released throughout the entire spring and summer, providing continuous primary infection.

ENVIRONMENTAL CONDITIONS FOR INFECTION

When the weather is moist, ascospores are produced and released throughout the entire spring and summer, providing continuous primary infection. The black rot fungus requires warm weather for optimal growth; cool weather slows its growth. A period of 2 to 3 days of rain, drizzle, or fog is also required for infection." "Conidia spores can also form, within cane lesions or on mummies that have remained within the trellis, and these are dispersed by splashing rain drops." Raindrops transfer these spores by moving the spores to different plant parts, especially susceptible young leaves. "If water is present, the conidia germinate in 10 to 15 hours and penetrate young tissue. New black rot infections continue into late spring and summer during prolonged periods of warm, rainy weather. The conidia are capable of germinating and causing infection several months after being formed. During August, the pycnidia are transformed into an overwintering stage that, in turn, gives rise to pseudothecia within which the spring spores (ascospores) are produced. This (ascospore) is "forcibly discharged into the air and can travel considerable distances." "Research has shown that ascospores are an important source of primary infections in the spring."In the spring during wet weather, the "pycnidia on infected tissues absorb water and conidia are squeezed out.""Conidia are splashed about randomly by rain and can infect any young tissue in less than 12 hours at temperatures between 60-90 degrees."A film of water on the vine surface is necessary for the infection to inoculate. This completes the disease cycle."

SYMPTOMS

Fig. Guignardia bidwellii 04

Relatively small, brown circular lesions develop on infected leaves and within a few days tiny black spherical fruiting bodies (pycnidia) protrude from them. Elongated black lesions on the petiole may eventually girdle these organs,causing the affected leaves to wilt.Shoot infection results in large black elliptical lesions. These lesions may contribute to breakage of shoots by wind, or in severe cases, may girdle and kill young shoots altogether

Infection of the fruit is the most serious phase of the disease and may result in substantial economic loss. Infected berries first appear light or chocolate brown, but quickly turn darker brown, with masses of black pycnidia developing on the surface. Finally, infected berries shrivel and turn into hard black raisin-like bodies that are called mummies.

CONTROL AND MANAGEMENT

A mixture of cultural and chemical control practices can manage grape black rot disease caused by *Guignardia bidwellii*. Cultural control aspects involve the basics in plant care and field sanitation as well as cleanup after an infectious outbreak. Chemical control has a large influence to eliminate disease.

Cultural control Cultural control consists of the management of the fields and sanitation methods to optimally grow grape crops.

Many universities and professors suggest the following cultural control practices:

The first cultural control method is to choose the right grape cultivar for the region that the grape will be grown in. Grape cultivars differ in their susceptibility to diseases, including differences in the disease black rot. Some varieties are less susceptible, while others are more prone to the disease when the right environmental conditions occur. Through research, many lists of grapes varieties will show the amount of resistance a variety has to disease and how the variety should be grown. A List of grape varieties is a good place to start for selecting the right grape variety.

Proper grape management in the field is important to control black rot disease. The placement of the grape plants is important to provide for proper aeration and lighting. "Choose a planting site where the vines will be exposed to full sun and good air circulation." Proper training of grape vines should be practiced. Many people use trellising to keep the vines off the ground. Trellising is a form of plant training, which is the arrangement of plant parts spatially to optimize productivity. Having the grape vines off the ground "reduces the time that vines remain wet from dew and rain and thus limits the amount of infection". As the current seasons growth occurs, tie the new growth in a spatial manner to allow for air circulation and sunlight to happen throughout the whole grape plant.

A proper pruning technique is another cultural control method to limit diseases. Prune each vine every year during the dormant period. This dormant pruning is heavily researched to provide the term-balanced pruning. This is done by using the weight of the previous year's cane growth (1- yr-old canes) to determine how many buds should be retained for the current production year. Up to 4 lbs of pruning can occur during the season. Remove excess growth, diseased and overwintering berries, leaves, and tendrils from the vineyard, and burn or otherwise destroy them. This practice reduces inoculum of the fungus, thus limiting disease. During the season minor pruning to train the canopy growth may occur to help with aeration.

The last cultural control method would be to maintain clean fields once infected. Keep the field well managed, do not allow overgrowth of weeds or plants near the grapes.

Once the plants have been infected "Remove all mummies from the canopy during the dormant pruning process. Mummies produce spores next to the susceptible grapevine tissues throughout the season; even relatively few can cause significant damage." Another technique to rid of infection can be to "cultivate the vineyard before bud-break to bury the mummified berries. Diseased berries covered with soil do not produce spores that will reach the developing vines".

Chemical control- The use of chemical control is widely available for agricultural purposes. To apply chemical applications, look at the fungicide label for proper use. Be sure that the conditions are optimal to spray to avoid drift and inefficiencies of the fungicide due to application. Please follow fungicide guidelines. There are a wide variety of chemicals that are available for both regular and organic growers.

Commercially, application of fungicides may be costly. To cut down on costs, we must understand the life cycle of the pathogens. Different fungicides are more effective at certain infection stages. Understand that throughout the development of the plant, different fungicides should be considered to protect plant health (Links for specifics with fungicides and plant development below). In order to cut down on spraying costs, it is important to understand life cycle of the pathogen. "Research in New York demonstrated berries of most varieties become resistant to black rot infection 3-4 weeks after bloom, therefore, sprays for black rot should not be needed at this time". Understanding times to limit application is important for good production practices. This shows that preventative chemical measures before the 3–4 weeks would be optimal. Each region should develop their own fungicide application program in correlation with cultural practices.

The most common fungicides that prove to be excellent for certain regions of the United States for controlling black rot are Sovran 50WG, Flint 50WG, Abound Flowable (2.08F), and Pristine 38WDG. Some of these fungicides may only be available for commercial use, however homeowners may go to a hardware store or fungicide distributor for similar fungicides.

Sovran 50WG is registered for control of black rot. The Sovran label gives different use rates for control of different diseases. For black rot the rate is 3.2-4.8 oz/A. Sovran is excellent for control of black rot.

Flint 50WG is in the same general class of chemistry as Abound and Sovran. It is registered for the control of black rot, powdery mildew and suppression of downy mildew. For black rot the rate is 2.0 oz/A.

Abound is in the same general class of chemistry as Sovran and Flint (strobilurin). Abound is recommended at the rate of 11-15.4 fl oz/A. In university tests, the rate of 11-12 fl oz provided good control of the mentioned diseases above. NOTE: Abound Flowable is very phototoxic to apples of the variety McIntosh or varieties related. Do not allow spray to drift from grapes to apples.

Pristine 38WDG contains a combination of two active ingredients (pyraclostrobin, 12.8% and boscalid 25.2%).Pyraclostrobin is in the same general class of chemistry as Abound, Sovran, and Flint (strobilurin Pristine is registered for use at the rates of 6-10.5 oz/A. A maximum of six applications may be made per season). The label states, "Do not use on Concord, Worden, Fredonia, or related varieties due to possible foliar injury. Experience in New York suggests that Corot noir and Noiret may also be sensitive to Pristine."

Black Rot Control for Organic Growers

Black rot is probably the "Achilles heel" for organic grape production in the East. In the only good trial that we've run with copper, it provided 40% disease control when applied at 2-week intervals, versus essentially 100% control with Nova. To try to avoid this, implement a rigorous program to remove mummies during pruning, and spray copper once a week throughout much of the growing season.

This was hard on some of the hybrid vines and runs counter to the thinking of many with a "sustainable" orientation (after all, copper is an element and doesn't break down into anything else, so it stays in the soil forever), but it did control the disease in an organically acceptable manner. Limit inoculum within the vineyard. Ideally, this would include removing or burying any mummies that you might encounter at the site. All mummified clusters should be removed from the trellis during pruning and from 2 to 6 weeks after cap fall and prune out any affected clusters before they allow the disease to spread, all the better (spores for disease spread are dispersed by rain primarily within the canopy, so should pose little risk of causing new infections if said clusters are simply dropped to the ground).

BOTRYTIS BUNCH ROT (*BOTRYTIS CINEREA*)

Botrytis bunch rot, caused by the fungus *Botrytis cinerea*, is a common problem wherever grapes are grown. The disease can cause serious losses in both yield and quality when weather conditions favour the disease.

SYMPTOMS:

Ripening grapes are affected by a rot which may progress to infect whole clusters. With sufficient rain and humidity, berries split open and develop a greyish mold on the surface. Affected berries may shrivel in the dry Okanagan climate.

Botrytis may cause girdling lesions on the pedicel or rachis, leading to drying of clusters or portions of clusters, and premature cluster drop.

Discoloured, shrivelled berries caused by Botrytis Bunch Rot. Photo courtesy P.Sholberg, Agriculture & AgriFood Canada.

INFECTION AND SPREAD:

Botrytis infects grape shoots, flowers, leaves and fruit under a range of temperature and humidity conditions (Table 1). Infection is optimal at 15-20 °C with free water or over 90% humidity. Grape cultivars with dense canopies, thin skins, and/or tight clusters are more susceptible to botrytis bunch rot (Table 2).

Botrytis overwinters as mycelium or sclerotia on bark and as mycelium in dormant buds. It may also overwinter in mummified fruit. In the spring, conidia (spores) are produced which can infect leaves and young clusters before bloom. The fungus may infect blossoms leading to fruit infection, but the fungus becomes inactive (latent) in the fruit possibly due to low sugar and high acid contents. It becomes active again when the berries begin to soften. Fruit infection usually begins in berries with 5 to 8 percent sugar (at veraison), with berries remaining susceptible up to harvest and also post-harvest. Fruit become increasingly susceptible has they mature.

Abundant conidia are produced on infected fruit, leading to secondary spread to adjacent berries in a cluster and to nearby clusters. Spread of the spores is aided by summer rains, overhead irrigation, heavy dew and juice from split berries. Spores are also carried by the wind. Mechanical damage caused by birds or insects creates entry points for the fungus and can greatly increase losses due to bunch rot.

Table. *Time/temperature relationships for botrytis bunch rot infection*

Temperature	Amount of Time
10 °C	30 hours or more
15.5 °C	18 hours or more
22.5 °C	15 hours or more
26.5 °C	22 hours or more
30 °C	35 hours or more

Slow drying conditions together with high humidity (90% or better) favour the development of *Botrytis*. Table 1 indicates the temperature required plus the corresponding period that slow drying conditions must exist for bunch rot infection to occur.

CULTURAL CONTROL OF BOTRYTIS BUNCH ROT:

Table. *Susceptibility of grape varieties in British Columbia to botrytis bunch rot*

Very Susceptible	Susceptible	Moderately Susceptible	Least Susceptible
Auxerrois	Barbera	Ortega	Cabernet Franc
Bacchus	Kerner	Scheurebe	Cabernet Sauvignon
Chardonnay	Lemberger	Schoenburger	Chancellor
Chasselas	Malbec	Syrah	Foch
Chenin Blanc	Muscat Ottonel	Vidal	Merlot
Ehrenfelser	Pearl of Csaba	Viognier	Petit Verdot
Gamay Noir	Pinot Gris		
Madeleine Angevine	Pinot Meunier		
Madeleine Sylvaner	Pinot Noir		
Optima	Sangiovese		
Pinot Blanc	Sauvignon Blanc		
Riesling			
Semillon			
Siegerebe			
Zinfandel			

During the growing season:

- Avoid overhead irrigation and keep irrigation periods as short as possible.
- Prevent excessive vine growth by judicious use of water and fertilizer.

- Canopy management, including shoot thinning and leaf removal before fruit set will reduce botrytis bunch rot. Remove leaves and lateral shoots located opposite, one node above, and one node below each fruit cluster.
- Prevent berry damage by effective control of powdery mildew early in the growing season.
- Minimize berry damage by birds and insects.

During the dormant season:

- Reduce carry-over of botrytis to the next season by removing any remaining fruit from the vines during pruning. If practical, remove trash from the vineyard floor.

CHEMICAL CONTROL OF BOTRYTIS BUNCH ROT:

Several fungicides are registered for the control or suppression of botrytis bunch rot. Alternate fungicides from different chemical groups to help prevent the development of resistance.

Table. ***Fungicides registered for bunch rot control on grape***

Fungicide	Chemical Group[1]	Active ingredients	Rate/ha	Rate/acre	PHI[2](days)	Comments
Rovral	2	iprodione 50%	1.5 kg/ha	600 g/acre	7	Maximum 2 sprays/season
Elevate	17	fenhexamid 50%	1.12 kg/ha	450 g/acre	7	Maximum 3 sprays/season
Pristine	7+11	pyraclostrobin 12.8% + boscalid 25.2%	420 – 735 g/ha	170-300 g/acre	14	Registered as suppression only for botrytis bunch rot. Maximum 2 sprays/season. Wine grapes only.
Luna Tranquility	7+9	fluopyram 125 g/L +pyrimethanil 375 g/L	1.2 L/ha	485 mL/acre	7	Maximum 2 sprays per season. Wine grapes only.
Scala	9	pyrimethanil 400g/L	2.0 L/ha	810 mL/acre	7	Maximum 3 sprays/season
Vangard	9	cyprodinil 75%	750 g/ha	300 g/acre	7	Maximum 2 sprays/season
Switch	9+12	cyprodinil 37.5% +fludioxonil 25%	775-975 g/ha	314-395 g/acre	7	Maximum 2 spays/season
Serenade MAX	NC	*Bacillus subtilis*	3-6 kg/ha	1.2-2.4 kg/acre	0	Biofungicide. Disease suppression only.

[1]Chemical Group: Products with the same number belong to the same class of compounds. Alternate products with different chemical groups to help delay or prevent the development of resistance.

[2]Pre-harvest Interval - time interval between last application and harvest

The number of spray applications necessary to control bunch rot depends upon disease pressure in the vineyard and weather conditions, as well as other factors such as susceptibility. Fewer applications may be needed if weather is very dry and/or disease pressure is low. Use a full program in vineyards where bunch rot was a serious problem in the previous year, and where sanitation measures were not taken after harvest. The number of sprays can be reduced in vineyards with a history of low disease levels and for all vineyards in dry years.

Key timings for *Botrytis* fungicide applications are bloom, just before bunch closing, veraison and pre-harvest. Botrytis control becomes more difficult as the grapevine matures because heavy canopy growth and bunch closing make it difficult to place the fungicide where it is needed.

Under typical Okanagan weather patterns, high risk periods for *Botrytis* infection are during bloom and from 1 to 2 weeks after veraison to harvest. These growth stages often coincide with rainy periods during June and September. Be cautious about adding spreaders or stickers to fungicide sprays, as some spray adjuvants have been shown to increase the development of bunch rot on grape clusters. It is believed that spray adjuvants remove the natural wax protection on grape berries.

POWDERY MILDEW (*UNCINULA NECATOR*)

*'Pinot noir' vines severely affected with powdery mildew. Photo courtesy P. Sholberg, Agriculture & Agri-Food Canada. Severe powdery mildew on Chancellor grape cluster. Photo courtesy P. Sholberg, Agriculture & Agri-Food Canada.*Powdery mildew, also known as oidium, is caused by the fungus

Uncinula necator. This fungus has a narrow host range attacking only grape plants and a few related species. It is the most common and widespread disease of grapevines in the Okanagan/Similkameen area. Popular wine grape varieties vary in susceptibility to powdery mildew (Table 1).

Powdery mildew on Reisling grape. Photo courtesy P. Sholberg, Agriculture & Agri-Food Canada. Early powdery mildew symptoms on Chancellor grapes. Photo courtesy P. Sholberg, Agriculture & Agri-Food Canada.

Table. Level of resistance of grape cultivars to powdery mildew

Susceptible	Intermediate	Least Susceptible
Bacchus	Chelois	Auxerrois
Cabernet Franc	Chenin Blanc	Malvoisie
Cabernet Sauvignon	Concord	Melon
Chancellor	Foch	Pinot Gris
Chardonnay	Pinot blanc	Semillon
Chasselas	Malbec	
Gamay	Merlot	
Gewurztraminer	Ortega	
Grenache	Pinot Noir	
Himrod	Perlett	
Madeleine Angevine	Sheridan	
Madeleine Sylvaner	Vidal Blanc	
Malbec	Weissburgunder	
Muller Thurgau		
Pearl of Csaba		
Petit Verdot		
Rkatzeteli		
Riesling		
Sauvignon blanc		
Schonburger		
Siegerebe		
Syrah		
Viognier		

SYMPTOMS

Powdery mildew symptoms can be seen on foliage, fruit, flower parts and canes. Mildew usually appears first as whitish or greenish-white powdery patches on the undersides of basal leaves. It may cause mottling or distortion of severely infected leaves, as well as leaf curling and withering. Lateral shoots are very susceptible. Infected blossoms may fail to set fruit. Berries are most susceptible to infection during the first three to four weeks after bloom, but shoots, petioles and other cluster parts are susceptible all season. Infected berries may develop a netlike pattern of russet, and may crack open and dry up or never ripen at all. Old infections appear as reddish brown areas on dormant canes.

Early powdery mildew infections can cause reduced berry size and reduced sugar content. Scarring and cracking of berries may be so severe as to make fruit unsuitable for any purpose. Be aware that many winemakers have a very low tolerance for powdery mildew on grapes. Research has shown that infection levels as low as 3% can taint the wine and give off-flavours.

LIFE CYCLE

The powdery mildew fungus overwinters as chasmothecia (tiny, round, black fruiting bodies), in bark, on canes, left-over fruit, and on leaves on the ground. Spores (ascospores) from the overwintering chasmothecia are released in the spring after a rainfall of at least 2.5 mm. For primary infection to occur the spores require at least 12-15 hours of continuous wetness at 10-15 °C to infect developing plant tissue.

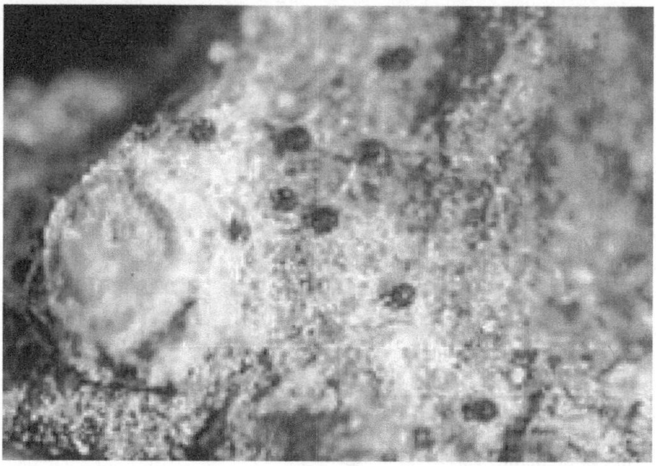

Once primary infection has occurred the disease switches to its secondary phase. Secondary colonies (white mildew patches) form in 7 to 10 days, although the disease is not noticeable early in the season. The white patches of powdery mildew produce millions of spores (conidia) which are spread by wind to cause more infections. Free moisture is not needed for secondary infection; temperature is the most important environmental factor (Table 2). The disease spreads quickly in early summer when temperatures are moderate. The incubation time (the time between infection and the production of spores) can be as short as 5 to 6 days under optimal temperatures. Shaded and sheltered locations favour mildew development. High temperatures and sunlight are inhibitory to powdery mildew. Extended periods of hot weather (>32°C) will slow the reproductive rate of grape powdery mildew, as well as reduce spore germination and infection.

Powdery mildew chasmothecia on grape shoot. Photo courtesy P. Sholberg, Agriculture & Agri-Food Canada. Magnified view of a chasmothecia and ascus containing ascospores. Photo courtesy P. Sholberg, Agriculture & Agri-Food Canada.

Severe powdery mildew infection on 'Chancellor' grape leaf, with developing chasmothecia. Photo courtesy P. Sholberg, Agriculture & Agri-Food Canada. Spore (condia) production of powdery mildew on a grape leaf. Photo courtesy P. Sholberg, Agriculture & Agri-Food Canada.

Powdery mildew infection on 'Chancellor' grape foliage. Note whitish mildew and purple discolouration. Photo courtesy P. Sholberg, Agriculture & Agri-Food Canada. Powdery mildew infection on 'Pinot Noir' grape cane. Photo courtesy P. Sholberg, Agriculture & Agri-Food Canada.

TABLE. Effect of temperature on the development of grape powdery mildew

Temperature of leaf* (°C)	Days for spores to develop and infect vine parts and produce new spores
6	32
9	25
12	18
15	11
17	7
23	6
26	5
30	6
33 (for at least 3 days)	0 (but 10% can recover in 5 days)
40.5 (for at least 6 hrs)	0 (kills the fungus)

* Note: leaf temperature can be slightly higher or lower than air temperature

DISEASE MANAGEMENT

Cultural Control:

1. Manage canopies to increase air drainage and light penetration by removing lateral shoots in dense canopies. If necessary remove leaves in the fruiting zone. Dense canopies provide low light intensity, which favours powdery mildew development;
2. Use an under-vine irrigation system (drip or micro-jet);
3. Manage irrigation carefully. Excessive irrigation leads to excessive vigour and higher disease potential.
4. Select varieties that are less susceptible to mildew (Table 1);

DISEASE FORECASTING:

Start mildew programs before the overwintering fungus can infect new growth. The first few treatments are the most important, and should be applied starting at budbreak or early shoot growth. Protective fungicide treatments prevent infection of grape tissue by fungal spores. Good coverage is important.

A mildew risk model can be used to forecast disease severity of secondary infections. The UC model developed at the University of California, Davis is the one most widely available and is sold with weather instrument software. The UC model requires a data logger for leaf wetness and temperature. Initially the model predicts primary infection based on hours of leaf wetness and temperature and then switches to the risk phase based only on temperature. The risk indices can be used to help time fungicide applications. When the risk is high the model recommends that fungicides be applied more often. In trials in California, the model has reduced the amount of fungicides applied to grapes. In tests in the Okanagan, the primary phase of the model has not been accurate and more research is required to adapt the model to local conditions. However the risk phase of the model could be valuable for assessing the risk of secondary infection during the growing season.

Daily analysis of the model allows the grower to visualize what the conidial population will be approximately one week later and what the potential disease severity will be two weeks later, allowing them to plan their fungicide program in terms of product and application interval.

Several years of data from many different sites around the Okanagan and Similkameen indicate that the risk mode consistently predicts severe powdery mildew and the shortest interval between fungicide sprays in July and August. Typical Okanagan temperatures during the summer months are typically optimum for powdery mildew.

For more information on powdery mildew forecasting models, refer to the University of California website at:http://www.ipm.ucdavis.edu/DISEASE/DATABASE/grapepowderymildew.html. If not using a disease forecasting model, apply powdery mildew control materials as frequently as necessary when severe mildew conditions exist, keeping in mind the leaf temperature and the number of days needed for spores to infect grape parts and produce new spores (Table 2).

TIMING FUNGICIDE SPRAYS:

Protect grape foliage from primary infection by application of fungicides from early shoot growth until after bloom. Good control early in the season to prevent establishment of the disease is the key to preventing a powdery mildew epidemic later in the summer.

Apply fungicides at the following growth stages.

1. When new growth is 5 to 10 cm long
2. Just before or immediately after bloom.
3. Every 10 to 14 days until grapes begin to soften and red varieties begin development of colour and white varieties change from green

to white or yellow. If sulphur is used, shorten the spray interval to 7-10 days.

Contact your winery at the beginning of the season to determine the acceptable pre-harvest intervals for any pesticides or sulphur products that may be used in the growing season. Some products contribute to the development of off-odours and off-flavours and may interfere with the fermentation process.

FUNGICIDE NOTES

Resistance Management

Grape powdery mildew has developed resistance or reduced sensitivity to sterol-inhibiting fungicides (Nova, Inspire) and to strobilurin fungicides (Flint, Sovran, Pristine) in other areas such as Eastern North America. Avoid over-using these products to prolong effectiveness in this area.

To help prevent resistance from developing:

- Alternate between different fungicide groups. Do not use more than 2 back-to-back sprays of fungicides with the same group number.
- Limit the number of sprays of products with a high risk of resistance (Nova, Inspire, Sovran, Flint, Pristine) to 2 per season per chemical group. Mildew fungicides with a low risk of resistance include Kumulus, sulphur, lime sulphur, PureSpray Oil, Milstop, Regalia, Actinovate, copper, and Serenade. Other products have a medium risk of resistance.
- Use only recommended dose rates.
- Ensure sprayer is properly calibrated to deliver accurate and thorough coverage.
- Integrate with non-chemical control methods.
- Discontinue use of a product if resistance is suspected and consult your crop advisor.

Sulphur is currently the most common fungicide used for powdery mildew control in the B.C. interior on both conventional and organic vineyards. Kumulus and Microthiol are dry flowable formulations of sulphur. The addition of 730 mL/ha of a wetting agent will help to improve the distribution of wettable sulphur over the plant surfaces. However, the use of these agents may cause more Botrytis infection. Too much wetting agent will cause excessive foaming in the sprayer. Fruit and leaf "burning" may occur if sulphur is applied during slow drying conditions or when temperatures are above 27°C. Do not apply sulphur to Concord, Sheridan or Foch.

PureSpray Green Spray Oil 13E is registered for suppression of powdery mildew on grape (summer spray), and provides protective, eradicant and antisporulant effects against grape powdery mildew. Do not apply in a spray volume of less than 1000 L water/ha (1% solution) or crop injury may result.

Use the shorter spray interval when disease conditions are severe. Oil will remove the bloom on grapes, so a longer PHI is recommended for table grapes. Do not use copper and oil together with fruit present. Do not use oil within 14 days before or after sulphur or captan fungicide.

It has been found in other areas that repeated sprays of oil can slightly reduce brix. It is suggested that the number of applications be limited to 3 per season, in early summer.

Quintec is a protectant fungicide, best used before visible mildew is present. Use the higher rate during high mildew pressure conditions. Rotate with fungicides from other chemical groups for resistance management. Quintec does not provide control of botrytis bunch rot.

Luna Tranquility is a premix containing 2 fungicides, fluopyram (group 7) + pyrimethanil (group 9). Controls both botrytis and powdery mildew. Apply preventatively. Rotate with other groups and do not use more than 3 times per season.

Vivando is registered for control of powdery mildew in grape. Rotate with fungicides from other chemical groups for resistance management, and do not use more than 2 times sequentially. Not effective on botrytis bunch rot.

Sovran and Flint are strobilurin fungicides (group 11) which provide good control of powdery mildews. These products are at high risk for resistance. Use the higher rate under heavy disease pressure.

Use 14-day intervals under moderate to heavy pressure, and 21-day intervals under low disease pressure. To help prevent the development of resistance, limit the use of group 11 fungicides to a maximum of 2 times per season. Alternate with at least 2 applications of fungicides in different chemical groups. Caution: Sovran drift may cause severe injury to certain cherry and Asian pear varieties.

Pristine is a premix containing 2 fungicides, pyraclostrobin (group 11) and boscalid (group 7). Begin applications prior to the onset of disease. Use a 10-14 day interval if using the low rate, or a 21 day interval if using the high rate. To help prevent the development of resistance, limit the use of Pristine (and other group 11 fungicides including Sovran and Flint) to a maximum of 2 times per season due to the high risk of resistance. Alternate with fungicides from different groups. Pristine also suppresses botrytis bunch rot.

Nova and Inspire (group 3 fungicides) are registered for powdery mildew control. Note that grape powdery mildew has developed a reduced sensitivity to this fungicide group in some areas of the world. To prevent the development of resistance, alternate group 3 fungicides with fungicides from other groups, and do not use more than 1-2 times per season. Do not apply Nova with copper fungicides such as fixed copper, copper oxychloride, copper sulphate or other copper containing products because this combination reduces the effectiveness of Nova.

Cantus (formerly known as Lance) may provide some suppression of bunch rot, in addition to control of powdery mildew. Apply on a 10-14 day schedule. Use the shorter interval when disease pressure is high. Rotate with other fungicides for resistance management.

Milstop is a contact fungicide that controls powdery mildew. It does not provide control of botrytis. It is considered a weaker product than sulphur, but may provide a good alternative for use on sulphur-sensitive cultivars. Apply using a sufficient volume of water to insure complete coverage of all stems and foliage. Use the high rate and short application interval (7 days) when conditions favour the development of powdery mildew. OMRI approved for organic production.

Serenade Max is a biofungicide derived from the beneficial bacterium *Bacillus subtilis*. It is registered for "suppression" of powdery mildew, and will also provide some suppression of bunch rot and sour rot. Use preventatively. OMRI approved for organic production.

Regalia Maxx is a biofungicide derived from a plant (giant knotweed). It is registered for suppression of both botrytis and powdery mildew. Mode of action is by host plant defence induction.

OMRI approved for organic production. Effectiveness on grape diseases is uncertain. Suggested use pattern: preventative application in rotation or tank mix with conventional fungicides. Avoid using as a solo product during periods of high susceptibility.

Bordeaux Mixture is a broad-spectrum fungicide and bactericide that is not easily washed off by rain.

Experience in Ontario indicates post harvest Bordeaux sprays using copper sulphate as a copper source to control late powdery mildew infections also appear to be an aid to increased bud hardiness and improved wood maturity in older vines. Sprays of fixed copper to young vines one month before the first fall frost appear to help "induce" some level of bud hardiness and wood maturity.

Copper sulphate (Bordeaux) is registered for several grape diseases including downy mildew, black rot and dead arm. It is not a strong powdery mildew fungicide, but may be useful for post-harvest control of light mildew problems, or as part of a mildew program on sulphur-sensitive varieties. Do not mix copper with other fungicides or insecticides.

Copper applied under cool, slow drying conditions may cause injury, even to so called tolerant varieties. Copper oxychloride is less injurious to plant tissue than copper sulphate. Addition of the lime helps to minimize, but may not prevent injury.

The label rate for copper sulphate on grape is 3 kg copper sulphate + 6 kg hydrated lime in 1,000 L water/ha. Also used is copper oxychloride (50% copper) at 3 kg + 6 kg hydrated lime in 1,000 litres of water.

Making Tank-Mix Bordeaux

1. Start water flowing into spray tank.
2. When tank is about one-third full and the mechanical agitator is in operation, start washing the powdered copper sulphate into the tank through a screen with water from the supply hose. A wooden spoon is often helpful in working powdered copper sulphate through the screen. Pre-soaking the copper sulphate in a plastic bucket of hot water will speed up the process.
3. By the time the tank is two-thirds full, all of the powdered copper sulphate should be in the tank. Then wash the lime (hydrated or builders') through the screen, using the water supply hose, into the copper sulphate tank. The lime should be as dilute as possible before it meets the copper sulphate solution in the tank, so use lots of water to wash lime through the screen. Pre-soaking the lime before adding to the tank may be preferred to washing powdered lime directly through the screen into the tank.
4. Keep the agitator running continuously and apply the Bordeaux Mixture immediately.

Dormant spray: Lime Sulphur is effective at suppressing the overwintering population of powdery mildew. It should be applied in the late fall after leaf drop or early spring before bud break to dormant vines to kill powdery mildew chasmothecia (initial inoculum). Good spray coverage of dormant vines is important.

SOUR ROT

Sour rot has not been a commonly recognized problem in the Okanagan/ Similkameen, but it was a serious issue in many vineyards during the 2004 growing season. Varieties which were affected include Pinot Blanc, Pinot Gris, Riesling, Pinot Noir and Chardonnay. It is likely that sour rot has always been present, but it may have been confused with botrytis bunch rot in the past. Sour rot affects both crop yield and wine quality. Infected fruit can give an unpleasant flavour to the finished wine and push volatile acid beyond acceptable levels.

SYMPTOMS

Symptoms include a soft watery rot with leakage of berry juice, and a distinctive vinegar smell. Large numbers of fruit flies, (also known as vinegar flies) and fruit fly larvae are generally present. The fruit flies produce ethyl acetate, a common fault in wine. There is very little tolerance of fruit contaminated with ethyl acetate. Sour rot appears similar to botrytis bunch rot. Affected berries are brick-coloured in white cultivars and purplish-brown in red cultivars.

Grapes infected with (as yet) unidentified yeasts, causing sour rot. Photo courtesy Agriculture and Agri-Food Canada

Sour rot in 'Optima' grapes. Photo courtesy Agriculture and Agri-Food Canada

DISEASE CYCLE

Sour rot is caused by a number of undesirable yeasts and bacteria, often in association with other fungal rot diseases including *Botrytis*, *Penicillium* and *Rhizopus*. Causal agents of the 2004 Okanagan outbreak of sour rot are under investigation.

Sour rot pathogens gain entry into grape berries through cracks and wounds caused by wasps, hail, birds, powdery mildew, botrytis, or by berry splitting caused by excessive vigour in tight clusters.

Such injuries attract fruit flies, which also effectively carry and spread the disease-causing organisms. During favourable conditions (warm moist weather, sugar accumulation in berries) the fruit fly will lay hundreds of eggs and start a new generation every ten to twelve days. Under the right conditions this can lead to explosive disease outbreaks. Fruit left on the ground from post-veraison bunch thinning can also be a contributing factor to build-up of fruit fly populations.

Cultivars with tight clusters and thin skins are generally more susceptible to sour rot. Very vigorous vines seem to be more prone to the disease due to tighter bunches with berries more likely to burst, and also due to a higher moisture environment.

The onset of hot, dry weather can lead to the drying of the infected and damaged berries, thereby slowing the spread of the disease. Cool weather will considerably slow fruit fly population growth, in turn leading to a reduction in the spread of the disease.

Some local and Californian observations indicate that the use of botrytis fungicides during periods favourable to sour rot development may increase damage caused by sour rot. Some experts speculate that the fungicide may kill the micro-flora present on the berry surface. These organisms may have some protective effect against the yeasts and bacteria responsible for sour rot. However, research done in Italy in the early eighties found that powdery mildew and botrytis control, together with fruit fly control, significantly reduced the incidence of sour rot.

Prevention and Control

- Reduce excessive vigour through the use of cover crops and/or the reduction of nitrogen fertilization and irrigation.
- Leaf removal between fruit set and veraison can lead to tougher berry skins, thereby reducing the risk of lesions.
- Control wasps through trapping and nest removal.
- Control powdery mildew to reduce grape berry lesions, which attract fruit flies.
- Prevent bird damage.
- Bunch thinning is best done before or during veraison. Berries removed during late season thinning have higher sugars which can contribute to the fruit fly population increase.
- Discard affected bunches before or during harvest to minimize the negative effects on the wine.
- No insecticides are registered for the control of fruit flies on grape. Fruit flies quickly become resistant to most chemical control products due to their fast rate of multiplication.
- Some French sources recommend 2-3 applications of Bordeaux mixture (copper sulphate and hydrated lime) at ten to twelve day intervals around the time of veraison. Copper applied to the fruiting zone has the effect of thickening the grape berry skin, hence reducing the risk of lesions.

CROWN GALL(*AGROBACTERIUM VITIS*)

Crown gall is an important disease of wine grapes when they are grown in cold climates. Incidence may range from a few vines in a vineyard to 100% of the vines. Gall development may result in girdling of vines and reduced vigour and yield.

SYMPTOMS

Gall formation on the aerial part of the vines is the most common symptom associated with this disease. The bacterium which causes crown gall may be present in plants that do not show any symptoms. Galls are usually noticed as swellings near the base of the vine and up the trunk.

Galls on roots of grape are not typical, however the bacteria can induce a localized necrosis of roots. Young galls are soft, creamy to greenish in colour, with no bark or covering. As they age, the tissue darkens to brown. The surface becomes open and the texture becomes moderately hard and very rough. The surface tissue of the galls turns black as it dies, but the bacterium remains alive in the vine.

Crown galls forming on 'Merlot' grape seedling. Photo courtesy Agriculture and Agri-Food Canada Aerial crown gall on grapevine. Photo courtesy Agriculture and Agri-Food Canada

DISEASE CYCLE

The crown gall bacterium survives systemically inside the grapevine. The primary source of diseased vines appears to be through propagation of diseased wood. The bacterium remains inside the vine until there is an injury to the trunk and only then invades the outer part of the trunk where it causes rapid cell multiplication and distortion of tissue producing galls.

Most grape cultivars are susceptible to crown gall. Major outbreaks are generally restricted to the varieties that are most susceptible to winter injury. The amount of crown gall present from year to year appears to be related to the severity of the preceding winter and the maturity of the vines.

The crown gall bacterium may also survive in vineyard soils in vine debris, but the importance of soil-borne inoculum in replanted vineyards is not well understood. It is believed that the majority of infections are a result of contaminated planting material.

CONTROL

Control of crown gall can at best be referred to as an attempt to manage this disease.

BEFORE PLANTING

Losses of grape plants due to crown gall may be minimized with some considerations before vineyard sites are selected and before vineyards are planted. Criteria to consider include the following:

- Select sites with good air and water drainage.
- Avoid planting in frost-prone areas.
- Select rootstocks that are resistant to crown gall. Certain rootstocks such as Courderc 3309, 101-14 Mgt, and Riparia Gloire are resistant, whereas Teleki 5C and 110 Richter are susceptible. Resistant rootstocks can reduce the amount of crown gall that appears on susceptible scions.
- Select hardy varieties where possible. In general, *V. vinifera* cultivars are all highly susceptible.

- Prepare sites with nutrients and lime before planting if necessary to avoid vine stress due to poor nutrition or low pH.
- Consider pest control programs for nematodes and phylloxera before planting through the use of soil fumigation or rootstocks. Feeding damage by these pests provides sites for the entry of crown gall bacteria.
- Plant old vineyard areas where crown gall was present only after grapevines have been removed for at least 2 years. This is important because crown gall bacteria can survive in the remnants of the old grape plants until the debris decomposes. Success in reducing crown gall from the soil by leaving soil fallow or rotating to other crops may vary depending on the amount of vine debris that is left in the soil, and how fast it breaks down.
- Do not propagate wood taken from galled vines.
- Specify that plant material you purchase be propagated from crown gall free plants. Producers must promote the production of certified indexed crown gall free vines.
- Hot water treatment of vines is effective in reducing crown gall infection levels in planting materials. Hot water treatment is required for vines imported from France or Germany for the prevention of phytoplasma diseases.

AFTER PLANTING

There is little that can be done to control this disease once it is established in the vineyard.

Avoid injury to vines (winter, mechanical and human). Remove suckers when shoots are small (3 to 6 cm) to reduce trunk damage and promote rapid healing of wounds. Removing larger shoots before they harden will result in clean and small scars.

- Hill young vines with 30 cm or more of soil or other material to protect them from cold temperatures. Remove this material carefully to avoid mechanical damage.
- Prune to delay budbreak (late pruning) on varieties prone to early bud break.
- In France, application of K2O instead of nitrogen fertilizers is used to improve resistance of the vines to cold.
- Biological control of grape crown gall has been effective in trials, and is a promising management tool for the future.

GRAPEVINE PHYTOPLASMA DISEASES

- Impacts
- Symptoms

- Disease Cycle
- Control
- Prevention
- Further Information

Phytoplasmas are microscopic plant pathogens, similar to bacteria, but much smaller and lacking cell walls. They live in the vascular system of plants and are spread by sap-feeding insects, including leafhoppers and planthoppers. There are several diseases of grapevine caused by phytoplasmas:

- Flavescence dorée
- Bois noir (black wood, vergilbungskrankheit)
- Grapevine yellows

Bois noir was detected in B.C. and Ontario in plants imported from France. The plants were removed and subsequent surveys have not detected new infections.

Flavescence dorée symptoms on Chardonnay, including downward leaf rolling and shrivelling of berry clusters. Photo courtesy INRA, France

Flavescence dorée is the most important and destructive phytoplasma disease of grapevines. Bois noir is less serious due to a slower rate of spread. Bois noir and flavescence dorée are found in several European countries. Phytoplasma diseases of grape elsewhere in the world are referred to as grapevine yellows, including an American grapevine yellows that occurs in New York and Virginia, and an Australian grapevine yellows.

IMPACTS

Flavescence dorée has been called "catastrophic" in France. Impacts include reduced vitality of vines, yield reductions, and reduced wine quality due to high acid and low sugar contents of fruit from infected plants. Without control measures, the disease spreads rapidly, affecting up to 80-100% of vines within a few years. Flavescence dorée has destroyed large viticultural areas in France and Corsica. It is still spreading despite mandatory control programs.

All *Vitis vinifera* varieties are susceptible to phytoplasmas. Popular cultivars such as Chardonnay, Cabernet Sauvignon, Pinot noir, Riesling, Sauvignon blanc, and Sémillon are highly susceptible to flavescence dorée. Some cultivars such as Sangiovese and Garganega are extremely susceptible and are killed quickly.

SYMPTOMS

All grapevine yellows diseases have similar symptoms, including growth reduction, leaf discolouration, downward rolling of leaves, and reduced quality and quantity of fruit. Symptoms are not uniform, and may appear on some or all shoots of infected vines. A few rootstock varieties are tolerant to flavescence dorée, and can be symptomless carriers of the disease.

Bois noir infected leaves on Sylvaner x Riesling in Mosel, Germany. Photo courtesy Clemson University, Plant Virus Slide Series; photo credit W. Gartel. Flavescence doréee infected Baco 22A grapevine, Armagnac, France. Photo courtesy Clemson University, Plant Virus Slide Series; photo credit W. Gartel.

Flavescence doréee symptoms on Cabernet Sauvignon, including shrivelled berries and reddish leaf discolouration. Photo courtesy INRA, France Bois noir symptoms on Pinot Noir. Photo courtesy INRA, France

Leaf symptoms in white cultivars appear as small, yellow spots along the main veins. These spots enlarge to form yellow bands along the veins, which gradually extend over large parts of the leaf. Red cultivars develop a similar pattern on the leaves, but the discolorations are reddish. Infected shoots often fail to lignify, and appear thin and rubbery. They later become brittle, sometimes with bud necrosis. Affected branches blacken and die during the winter. Numerous small black pustules form along the diseased branches of susceptible cultivars.

Fruit set is reduced on grapevines infected early in the season, as the inflorescences dry out and fall off. In later infections, bunches become brown and shriveled. Premature berry drop occurs in some cultivars.

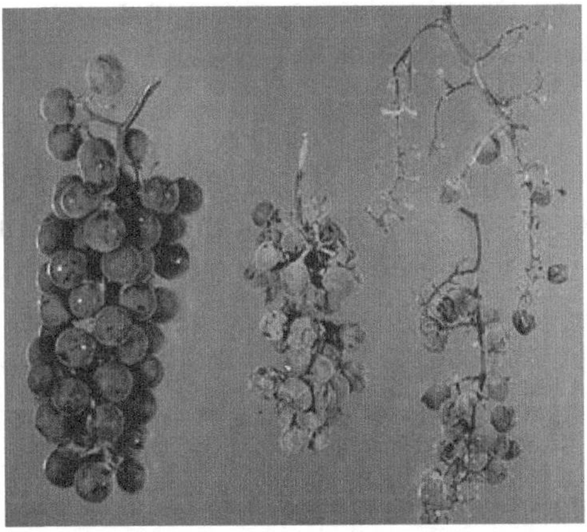

American grape vine yellows - healthy and diseased clusters of DeChaunac. Photo courtesy Clemson University, Plant Virus Slide Series; photo credit J.K.

Uyemoto.Bois noir - black pustules on shoot. Photo courtesy Clemson University, Plant Virus Slide Series; photo credit W. Gartel.

DISEASE CYCLE

Flavescence dorée is spread by the leafhopper *Scaphoideus titanus*, which is native to eastern North America. It is known to occur in Canada from the Maritimes through to Saskatchewan, but it has not yet been detected in British Columbia.

S. titanus spends its whole life cycle on grapevines. It is a highly mobile and efficient vector that is largely responsible for the epidemic spread of flavescence dorée. Both nymphs and adults are able to acquire the phytoplasma while feeding. After a latent period they are able to transmit the disease until they die. This leafhopper overwinters as eggs which are inserted (laid) into the bark of grapevines.

Bois noir is transmitted by a few planthopper vectors, the main one being *Hyalesthes obsoletus*. This insect is not known to be in Canada. It prefers herbaceous weeds over grapevines and transmits the phytoplasma from these weeds to grapevines. Bindweed (*Convulvulus arvensis*), nettle (*Urtica dioica*) and hedge bindweed (*Calystegia sepium*) are the main reservoirs in Europe. Since the vector only feeds incidentally on grape, bois noir is a much slower spreading disease than flavescence dorée. It is widespread in many parts of Europe. The recent identification of new bois noir vectors in Europe raises concerns regarding related insect species in Canada.

Both phytoplasmas and leafhoppers can be spread with propagation material. Flavescence dorée is symptomless in some cultivars, and it also has a long (up to 3 year) latent period before symptoms can be seen. Thus even apparently healthy vines may be carrying grapevine yellows diseases. A few rootstock varieties are symptomless carriers that have been responsible for spread of flavescence dorée in Europe.

CONTROL

Control of phytoplasma diseases involves scouting and removal of infected or symptomatic vines, as well as control of the vector with insecticide sprays.

In France, flavescence dorée is a regulated disease. Compulsory control measures for both the disease and the vector were strengthened in 2003. When an infected grapevine is found, a quarantine area is delimited around it for a minimum of 2 years. All infected plants must be destroyed, and control of the vector with several insecticide sprays is required in that zone.

PREVENTION

Use only planting material from reliable, approved sources and produced in areas where phytoplasma diseases are not present. Canada has grapevine import regulations that help to protect grape growers from the introduction and spread of diseases and other pests which are not yet established in British Columbia. Selected grapevine variety and rootstock clones from Canadian-approved nurseries in France and Germany are currently approved for importation into Canada. Under this program, nursery plants are propagated from mother plants that were tested for regulated pests and diseases by the exporting country then grown under conditions to reduce the risk of contamination. The CFIA has tested samples of many of these grapevine varieties or rootstocks prior to approving their importation.

YOUNG VINE DECLINE

Decline of young vines may have several causes, including diseases, nematodes, environmental damage such as winter injury, and cultural factors. Often more than one factor or more than one disease may be involved.

OKANAGAN SURVEY RESULTS (2007-2008)

A recent survey conducted by Agriculture and Agri-Food Canada investigated decline problems in Okanagan vineyards (O'Gorman, Haag & Sholberg). The survey confirmed the presence of several fungal pathogens causing vine decline symptoms. Diseases detected included:

- Black foot disease (*Cylindrocarpon* spp.) - isolated from vines ranging from 3-15 years of age in several vineyards. Infection was associated with both a gradual and a rapid decline of vines.
- Esca (*Phaeomoniella chlamydospora* and *Phaeoacremonium aleophilum*) - detected in necrotic vascular tissue on young vines up to 6 years old. Both *Phaeoacremonium* and *Cylindrocarpon* were recovered from vines in one vineyard where over 50% of the vines showed decline symptoms.
- Botryosphaeria canker (*Botryosphaeria parva* and *B. dothidea*) - detected in vines ranging from 3-11 years of age in several vineyards. The vineyards where *B. parva* was isolated showed severe decline problems.
- Roesleria root rot (*Roesleria subterainia*) - found in several vineyards in vines that were also infected with*Cylindrocarpon*.

Note that black foot, esca and botryosphaeria canker are all new diseases that have not been previously diagnosed in British Columbia vineyards.

GENERAL SYMPTOMS

General symptoms of grapevine declines caused by fungal pathogens include delayed and stunted growth, short internodes, yellowing and premature loss of leaves, tendril dieback, trunk dieback, dead arm and cankers. Discolouration may be observed in the wood when vines are cut open. Decline may be rapid, causing plant death within 2-3 weeks, or slow, resulting in reduced vigour and yield over a period of years.

DIAGNOSIS

Vine decline symptoms can be difficult to diagnose accurately. B.C. grape growers interested in diagnosis of decline problems should contact the provincial plant diagnostic laboratory or PARC Summerland for more information.

ESCA / YOUNG VINE DECLINE

Young Vine Decline is a disease complex with various names and manifestations, including black measles, esca, young esca, Petri disease, apoplexy, and black goo disease. It has been an increasing problem in California vineyards, causing significant economic losses due to vineyard replanting and yield reductions. It has also increased in importance in Europe, most likely due to a reduction in the use of sodium arsenite fungicides, which have been banned in most countries.

In California, esca or black measles has been a problem in older vineyards for decades. More recently it has also become a problem in young vines, following a massive replanting effort which replaced phylloxera-susceptible rootstocks with other more resistant hybrid rootstocks. Unfortunately, the new rootstocks appear to be more susceptible to vine decline.

A 2007-2008 survey conducted by Agriculture and Agri-Food Canada investigated decline problems in Okanagan vineyards (O'Gorman, Haag & Sholberg). The survey confirmed the presence of several fungal pathogens causing vine decline symptoms, including 2 pathogens associated with esca.

Young vine decline in an Okanagan vineyard. Photo courtesy Agriculture and Agri-Food Canada

CAUSAL AGENTS

Esca and young vine decline are caused by several fungi which act alone, in combination or in succession to cause disease. Fungal pathogens involved in young vine decline in both B.C. and California include:

- *Phaeomoniella chlamydospora*
- *Phaeoacremonium aleophilum*

Other species of fungi have also been associated with esca vine decline in California, but have not yet been detected in B.C., including:

- *Phaeoacremonium inflatipes*
- *Phaeoacremonium chlamydosporum*

Several wood rotting basidiomycete fungi including a *Phellinus* species (now known as *Fomitiporia punctatus*) and *Stereum hirsutum* were historically associated with esca. While these fungi could be isolated from diseased vines, it was not possible to reproduce the same symptoms by inoculating healthy vines. It is now thought that infection with *Phaeoacremonium* and *Phaeomoniella* species predisposes the vines to wood rots by breaking down toxic phenolics and other host-produced substances which are inhibitory to wood rot fungi.

There are also several other diseases which cause decline of grapevines including Cylindrocarpon black foot, Botryosphaeria canker, Phomopsis canker (not yet found in B.C.) and several root rot diseases.

SYMPTOMS - ESCA ON YOUNG VINES

Phaeomoniella and *Phaeoacremonium* infect the vascular tissue of young vines resulting in darkening and plugging of the xylem vessels. Symptoms range from mild to severe, and include chlorotic foliage, defoliation, stunting of growth and reduced yield and quality. Sudden wilting and death (apoplexy) of vines or cordons may occur in midseason, particularly during hot weather. Vine symptoms also include delayed and weak growth in spring, and reduced lignification of canes in the fall. Discolouration in the xylem vessels may appear as black streaks when vines are cut longitudinally or as black dots when cut in cross section. Wood rot is usually absent on young vines.

Foliar symptoms include light green or chlorotic spots between the leaf veins or along the leaf margins. The spots gradually expand and turn yellow brown or red brown, leaving only a narrow strip of green tissue along the main veins. In California, a mosaic has also been observed on terminal leaves followed by shoot tip dieback. It is thought that foliar symptoms are caused at least partially by toxins produced by the fungi and/or reaction products of the infected wood which are translocated to the leaves.

Symptoms on berries may be present without symptoms on the leaves and vice versa. Berries on affected plants tend to be small, and may not reach full maturity. Sugar content and flavour can be adversely affected. Berries sometimes develop spotting or black measles symptoms, with white cultivars being more susceptible. Berry spotting consists of brownish to purplish, small pinprick sized spots on the berry skin. Heavy spotting may lead to skin cracking, berry shrivel, and secondary berry rots.

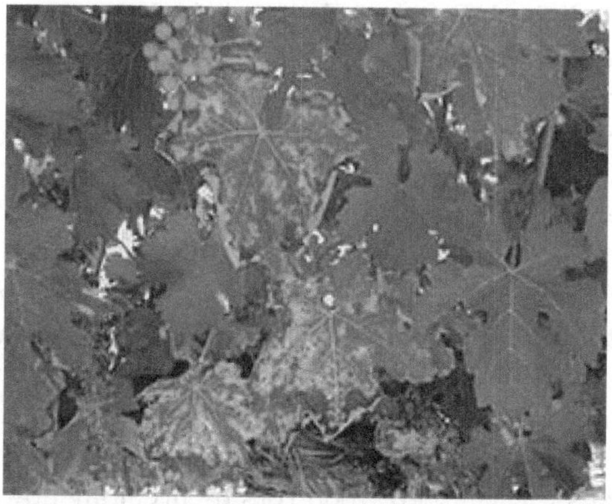

Esca symptoms on leaves. Photo courtesy W.D. Gubler, University of California. Black measles symptoms on berries. Photo courtesy W.D. Gubler, University of California.

SYMPTOMS - ESCA / BLACK MEASLES ON OLDER VINES

Esca is known as a disease of adult plants (8-10 years and older). It is caused by the same fungi that cause young vine decline. However on older vines there are often wood rot pathogens involved as well. Symptoms are similar to those described on young vines. It causes a decline with characteristic leaf and fruit symptoms, and sudden wilting and death in severe cases.

Additional symptoms in older wood include an internal white heart rot which causes the wood to become soft and spongy. A dark line or a series of black dots generally surrounds the white heart rot. In longitudinal section, this vascular discolouration appears as black streaks.

Vascular discolouration in wood appears as black dots in cross section or streaks in longitudinal section.

DISEASE CYCLE

Research underway in California is increasing our understanding of the biology of the fungi causing young vine decline. The method of disease spread is not yet certain, however spore trapping results from infected vineyards have found air-borne spores, with large numbers of spores trapped during rainfall. Infection most likely occurs through wounds.

Wet weather and warm temperatures appear to favour the disease. Years with above normal rainfall and high summer temperatures have resulted in more severe symptoms in California.

The wood rot fungi that are associated with esca are wound invaders, which infect through large wounds or pruning cuts on the trunk or cordons of mature vines. These fungi have a wide host range, and are commonly found as wood rots in other broad leaf tree species. Air-borne basidiospores produced by fruiting bodies (bracket fungi) on dead wood or other hosts can be spread long distances by wind, and can infect grapevines. Observed spread along rows in vineyards may also indicate that spores can be spread by tools during pruning and grafting.

Dissemination in Nursery Stock: The presence of vine decline affecting a high proportion of plants in recently established vineyards in California suggests that nursery stock may be carrying the disease. The pathogens have also been isolated from a high proportion of vine cuttings produced in Italy or France.

Prevention and Management:

- Plant healthy vines with no sign of discolouration in the vascular tissue.
- Propagation material should be taken from healthy vines with no wood darkening.
- Burn prunings, or remove them from the vineyard.
- Prune healthy-looking vines before sick vines. Disinfect pruning tools periodically, particularly after pruning vines with symptoms.
- The type of vine training system can be a factor. Extensive pruning cuts create more potential infection sites. One study found 0.1% esca in lateral cordon vineyards, compared to 15-20% in vines trained by double Guyot.
- There are no fungicide treatments for esca/young vine decline in Canada.

BLACK FOOT DISEASE (*CYLINDROCARPON* SPP.)

A survey of grape decline problems in Okanagan vineyards was conducted in 2007/08 by Agriculture and Agri-Food Canada (O'Gorman, Haag & Sholberg). The causal agent of black foot disease (*Cylindrocarpon* spp.) was isolated from vines ranging from 3-15 years of age in several vineyards. Infection was associated with both a gradual and a rapid decline of vines. This disease had not been previously diagnosed in grapevine from British Columbia.

Cylindrocarpon is known to occur in all major viticulture regions of the world, but the disease is not well understood. It's importance in B.C.'s vineyards remains to be seen. The following information is based on information from California and other viticultural areas.

SYMPTOMS

In California, *Cylindrocarpon* causes "black foot disease" in grape, affecting mainly young vines up to 8 years old. Roots show black, sunken, necrotic lesions. In cross section, the base of the trunk appears necrotic and xylem vessels may be black in colour. Leaves may appear to be water stressed or scorched, and vines may be stunted and/or killed.*Cylindrocarpon* may occur in combination with other plant pathogens.

Vine decline caused by Cylindrocarpon. Photo courtesy Agriculture and Agri-Food Canada Internal discolouration of vine rootstock infected with Cylindrocarpon. Photo courtesy Agriculture and Agri-Food Canada

DISEASE CYCLE

Cylindrocarpon is a common soil-borne fungus which causes root rot in many plant species. It survives in soil as mycelium and also produces conidia (spores) and chlamydospores (spores that can survive adverse conditions). Research in South Africa has shown that nursery soils can be a source of inoculum, and the disease can be carried on nursery plants.

DISEASE PREVENTION AND MANAGEMENT

- Plant healthy vines with no sign of root deterioration or discolouration in the vascular tissue.
- Hot water treatment has been shown in South Africa to reduce the level of *Cylindrocarpon* infection in nursery stock. Hot water treatment at a similar temperature and duration is required for vines imported from France or Germany for the prevention of phytoplasma diseases.
- There are no fungicide treatments for *Cylindrocarpon* available in Canada.

BOTRYOSPHAERIA CANKER

(*Botryosphaeria parva* and *B. dothidea*)

A survey of grape decline problems in Okanagan vineyards was conducted in 2007/08 by Agriculture and Agri-Food Canada (O'Gorman, Haag & Sholberg). Botryosphaeria canker (also known as bot canker and black dead arm) was detected in vines ranging from 3-11 years of age in several vineyards. Two species of *Botrytosphaeria* were identified, including *B. parva* and *B. dothidea*. The vineyards where *B. parva* was isolated showed severe decline problems. This disease had not been previously diagnosed in grapevine from British Columbia.

Bot canker is known to occur in most viticulture regions of the world. In California, recent research has shown that over 9 species of Botryosphaeria can infect grapevines, and that the disease is very prevalent in that state. The following information is based on information from California and other viticultural areas.

SYMPTOMS

Botryosphaeria produces black streaks in the xylem of infected wood, typically causing wedge-shaped cankers in the trunk and cordons. The cankers are most easily seen when the vine is cut in cross section.

Black pycnidia (fungal fruiting bodies) form on diseased vine parts under the bark of cordons, trunks, and spurs. Other diseases, including Eutypa canker cause very similar symptoms. Eutypa has not yet been identified in B.C. vineyards.

In California, botryosphaeria canker causes death of arms, cordons and vines, often so suddenly that there are no foliar symptoms. It is commonly seen in vines 10 or more years old. In other areas, foliar symptoms such as chlorosis and stunting are more common. Symptoms may vary depending on the species of Botryosphaeria present. Typically diseased vines fail to break dormancy or suddenly wilt during the following growing season.

Vine decline caused by Botryosphaeria canker. Photo courtesy Agriculture and Agri-Food Canada Internal discolouration of Botrytosphaeria-infected vines. Photo courtesy Agriculture and Agri-Food Canada

DISEASE CYCLE

The fungus overwinters as pycnidia on diseased wood. Pycnidia release conidia (spores) when wet by rain or irrigation. Conidia are spread by water splash and wind to cause new infections, primarily through fresh pruning wounds. Moisture is needed for infection, along with temperatures above 5 ° C.

Botryosphaeria can also infect other tree species, and could also be present in adjacent orchards, forests and riparian areas. The importance of spread from other host species to vineyards is not well understood.

Disease Prevention and Management:

- Remove visible cankers and dead or declining arms, cordons and vines.
- Burn prunings from infected vines, or remove them from the vineyard.
- Prune healthy-looking vines before sick vines. Disinfect pruning tools periodically, particularly after pruning vines with symptoms.
- In California, late pruning during March results in less infection of pruning wounds. Optimal timing may be different in our climate. In general, warmer temperatures accelerate healing of pruning wounds, so wounds are susceptible for a shorter time. Pruning during a period of forecast dry weather would also help to reduce infection.
- Good cultural practices help to maintain vigour and reduce losses.
- There are no fungicide treatments for Botryosphaeria canker available in Canada.

GRAPE ROOT ROT AND DECLINE

PHYTOPHTHORA AND PYTHIUM ROOT ROTS (*PHYTOPHTHORA* SPP., *PYTHIUM ULTIMUM*)

Root diseases caused by *Phytophthora* or *Pythium* spp. have not been common in the Okanagan/Similkameen in recent years. These diseases are typically sporadic, usually occurring on single vines that are excessively irrigated. Affected vines are smaller than nearby healthy plants and the foliage may become chlorotic or colour prematurely. A canker develops near the soil line and usually extends downward to the roots. Removal of the periderm exposes the necrotic woody tissue underneath. Roots may become blackened and decayed; and smaller fibrous roots may die. Death of the vines may occur within 2 years from the time symptoms appear.

ROESLERIA ROOT ROT

A root rot caused by the fungus *Roesleria subterranea* has also occurred sporadically in Okanagan vineyards. It is considered to be a weak root pathogen that colonizes injured or dead roots, from where it may also invade healthy roots. During a 2007/08 survey of grapevine decline problems, roesleria root rot was found in several vineyards in vines that were also infected with *Cylindrocarpon.*

Fruiting bodies of Roesleria subterranea growing on a grape root.Photo courtesy Michael Weis, Agriculture & Agri-Food Canada

CONTROL

Cultural: Prevent root rots with good water management and by providing drainage if needed. *Phytophthora* and *Pythium* are "water molds" which are favoured by very wet conditions.

In replant sites, remove as many old roots as possible before replanting vineyards.

Chemical: There are no fungicides registered for control of these soil-borne diseases on grape in Canada.

GRAPE NEMATODE PROBLEMS

Nematodes are translucent, microscopic roundworms, typically measuring 0.25 to 1 mm in length and only about 0.1 to 0.2 mm in diameter. Most nematodes in soil are beneficial components of the soil ecosystem, where they feed on bacteria, fungi and other micro-invertebrates, stimulating and regulating the turnover of nutrients. Their abundance ranges from about 1000 per liter of degraded or infertile soil, up to 50,000 per liter of highly fertile soil. Some nematode species, however, parasitize plant roots and are economically important plant pathogens.

SYMPTOMS

All plant-parasitic nematodes use hollow spear-like mouthparts called "stylets" to feed on plant roots. A few nematode species do not cause much direct damage on their own, but can transmit viruses while they are feeding. Most species cause direct damage, including gall-like malformations that impair root function or dieback of root tips and formation of lesions on the roots. When nematode population densities are large, root systems will be sparse, there will be few fine roots, and the damaged roots will appear reddish, brown or black rather than a healthy white colour. If root-knot nematodes are present, it may be possible to see characteristic swellings or galls on feeder roots. Vines severely damaged by plant-parasitic nematodes are stunted and have reduced yields.

At the field level, nematode damage is often overlooked because the above-ground symptoms are non-specific and difficult to distinguish from effects of other factors. Nematode populations usually exist in patches of high population densities; rarely, entire fields will be uniformly affected. Thus typical symptoms include poor vine vigour in patches of one to dozens of vines. This patchy, uneven growth caused by plant-parasitic nematodes may occur within a vineyard with uniform soil conditions. Areas with high nematode population densities may also be correlated with other soil factors that can affect vine vigour. For example, many species of plant-parasitic nematodes reach higher populations and cause more damage in sandy soils than in finer-textured soils. As a result, poor vine growth in sandy soils can be the result of low water or nutrient availability as well as the interaction of these additional stresses with nematode damage.

NEMATODE SPECIES

The main nematode species of significance to winegrape production in British Columbia are ring nematodes (*Mesocriconema xenoplax*), dagger nematodes (*Xiphinema* species), root-knot nematodes (*Meloidogyne hapla*) and root-lesion nematodes (*Pratylenchus penetrans*). Vineyards in British Columbia often have more than one species.

Ring nematodes (*Mesocriconema xenoplax*): These nematodes are ectoparasites that use long stylets to extract the contents of root cells, especially at root tips. They cause necrosis or blackening of roots like root-lesion nematodes, but their preferential feeding at root tips also causes stunting and sometimes formation of swollen or malformed root tips without obvious necrosis. Preferred hosts for ring nematodes include cherry, peach, apricot, and plum in addition to grape. These nematodes are known to cause significant damage to grape in California and Oregon. Ring nematodes have recently been found to be relatively widespread in the Okanagan Valley, and preliminary research at the Summerland Research Centre indicates that they can cause significant growth reduction of young grapevines, particularly self-rooted vines, under Okanagan growing conditions.

Root-knot nematodes (*Meloidogyne hapla*): These nematodes are endoparasites; infective juveniles invade root tips and establish permanent feeding sites. As the nematodes feed and mature in the root, galls or "knots" of root tissue form around them. These galls, which typically range in size from 3 to 12 mm in diameter, interfere with water and nutrient uptake. Several species of root-knot nematodes are recognized to be significant pests of grape in most grape-growing regions. To date, only one species has been found in B.C. vineyards, *Meloidogyne hapla*. The damage caused by*M. hapla* has not been studied as extensively as other species, and its impact on grapevine health under Okanagan growing conditions is not clear.

Dagger nematodes (*Xiphinema* species): *Xiphinema americanum* and several closely related species of *Xiphinema*are commonly found in B.C. vineyards. These nematodes are ectoparasites; they do not burrow into roots, but insert their long stylets deep into root tips, where they feed on root tip cells. Dagger nematode feeding causes limited necrosis and stunting and swelling of root tips, but they seldom cause significant damage on their own. Dagger nematodes are mostly important as vectors of viruses; *X. americanum* and related species transmit *Tomato ringspot virus*, *Tobacco ringspot virus*, *Peach rosette mosaic virus* and several other minor viruses.

Xiphinema index transmits the devastating *Grapevine fanleaf virus*. This nematode species can also cause significant direct damage in the absence of the virus. Plant protection measures that require treatment of imported planting stock have prevented *Xiphinema index* from being introduced into British Columbia.

Root-lesion nematodes (*Pratylenchus* species): These nematodes are migratory endoparasites, burrowing into roots to feed, and migrating freely between root tissue and soil. Root-lesion nematodes cause symptoms that range from many small lesions on young feeder roots to abnormal darkening and necrosis or death of the roots. Severely damaged root systems will have few newly formed feeder roots and sparse tufts of fine roots that will be blackened and crumbly. Most research documenting impacts of root-lesion nematodes has been conducted on *Pratylenchus vulnus*, a species that does not occur in B.C. The species most commonly found in the Okanagan Valley is *P. penetrans*. While this nematode clearly damages roots, the relationship between *P. penetrans* population densities and overall vine vigour is not well understood.

PREVENTION AND CONTROL

Nematodes are best managed before planting. Prevention involves testing for the presence of pathogenic nematodes followed by pre-plant soil fumigation to reduce nematode populations. There are no chemical treatments for nematodes that can be used on established vines in Canada.

NEMATODE RESISTANT ROOTSTOCKS

Rootstocks that are tolerant or resistant to plant-parasitic nematodes are available. These rootstocks have been evaluated against nematode species found in California and Mediterranean regions of Europe. Consequently, it is not clear if they would express the same tolerance or resistance to nematode species found in B.C. Rootstocks reported to be tolerant or resistant include: Dogridge, Harmony, Freedom, Ramsey, Teleki 5C, K51-32, and 1613C. Two other rootstocks, 420A Mgt and 101-14 Mgt, are reported to be resistant to ring nematodes, which are overall probably the most important nematodes in B.C. vineyards.

PRE-PLANT SOIL FUMIGATION

Soil fumigation with chemicals such as Vapam (metam-sodium) and Basamid (dazomet) are an option for reducing the population of pathogenic nematodes in soil prior to planting. They will also help to reduce the levels of soil-borne plant pathogens and viable weed seeds. Soil must be well prepared before fumigation, and should be moist. As many old grape roots should be removed as possible. These products release toxic gases on contact with moist soil. Following application, the soil surface should be sealed by watering and rolling or with polyethylene sheets. Refer to the product labels for detailed information on application and safety precautions.

Soil fumigation is expensive. Before considering, have samples of your soil analyzed for nematodes.

ALTERNATIVES TO FUMIGATION

Soil solarization is a non-chemical technique that will help to reduce the population of nematodes and other soil-borne pests. Solarization involves capturing the heat of the sun by covering the soil with transparent polyethylene plastic sheets during warm sunny months. The soil temperatures under the plastic increase to levels lethal to many soil-borne plant pathogens, weed seeds, seedlings, and nematodes. It is usually necessary to take the land to be solarized out of production for a year. Soil should be tilled before solarization, and should also have a good soil moisture level. The area to be treated should be free of weeds, plant debris, and large clods which would raise the plastic off the ground. Cover the area with a double layer of clear polyethylene sheet, seal the edges with soil and leave it in place for 4-6 weeks during the heat of the summer (mid-June through mid-August). If possible, leave the poly in place until replanting to prevent re-contamination. Black plastic is less effective than clear plastic.

Other cultural management techniques include green manuring, and/or rotating crops to reduce the numbers of pathogenic nematodes. These tactics do not guarantee a reduction in nematodes, as populations may be sustained on other species of plants including weeds.

NEMATODE TESTING:

Consider testing your soil for plant-parasitic nematodes before planting new vineyard blocks. If damaging populations are present, consider soil fumigation before planting.

Sampling strategies and interpretation of the results depend on the nematode species suspected and environmental conditions, and should be developed in consultation with a diagnostic laboratory. Populations of all nematode species vary seasonally. Since some nematode species migrate between roots and soil, it is usually necessary to collect both soil and root samples for testing. The laboratory analysis includes extraction of nematodes from soil and plant tissue, identification of the species present, and enumeration. The absence of recognized nematode pathogens in diagnostic samples can be used to rule-out nematodes as the cause of poor crop growth.

6

Grape Insect and Mite Pests

LEAFHOPPERS

In addition to the widespread Virginia creeper leafhopper (VCLH), the western grape leafhopper (WGLH) occurs on the east side of the Okanagan Valley from the north end of Penticton south to the U.S. border. The WGLH is more tolerant to insecticides and it is important to determine if it is present in your vineyard.

Western grape leafhopper adult Virginia creeper leafhopper adult

Leafhoppers are often distributed unevenly in vineyards and monitoring will help determine which areas require treatment. Unnecessary sprays reduce numbers of beneficial insects and spiders, resulting in a rapid resurgence in pest numbers and outbreaks of thrips, mites, and mealybugs. Experience has shown that WGLH numbers are low in organic vineyards; damaging populations in conventional vineyards generally result from repeated sprays of insecticides that do not control this species but damage numbers of the egg parasite, *Anagrus erythroneurae*. Successful control of leafhoppers relies on proper monitoring of numbers throughout the vineyard and the judicious use of insecticides in order to preserve beneficial insects and spiders.

IDENTIFICATION

Adults of both species are similar in appearance, nearly ½ cm in length with reddish-brown markings on a pale white or yellowish background. Adult WGLH can be distinguished by their pale eyes and irregular reddish-orange markings on a whitish background. VCLH have reddish-brown eyes and a more regular reddish-brown zigzag pattern on the wings. The background body colour is pale yellow to light brownish-yellow, producing an overall darker appearance. Adult leafhoppers are best identified on yellow sticky cards placed in outer vineyard rows during late April and May that can be inspected with the aid of a hand lens or magnifying glass.

Immature stages (nymphs) of WGLH are distinguished by their pale white colour, lack of pigmentation in the eyes, and the presence of one to three pairs of pale indistinct yellow spots on the thoracic segments of larger nymphs. VCLH nymphs have reddish-brown eyes and a pale yellowish body colour. Larger VCLH nymphs develop a pair of dark reddish-brown spots on the first body segment behind the head and a pair of reddish-orange spots on the other two thoracic segments. Nymphs can be identified feeding on the undersides of fully expanded leaves that show signs of feeding damage (stippling).

LIFE CYCLE AND DAMAGE

The biology and life cycles of the two species are similar. There are two generations each year. Adults of both species spend the winter in leaf litter or under plant debris in the vineyard or in nearby vegetation. Adults emerge on warm spring days to feed on a wide variety of plants, moving to grape, Virginia creeper and Boston ivy when the first leaves appear.

The overwintered females deposit eggs in the leaf tissue on the undersides of fully expanded leaves during May and June. The small, flattened opaque eggs are deposited side by side, usually in small groups of around two to six for the VCLH and singly for the WGLH. Nymphs of this 1st generation hatch mainly from mid June to the end of July. There are five nymphal stages. Winged adults that appear during July and August lay eggs that develop throughout the fall into overwintering adults. The WGLH develops more slowly than the VCLH and nymphs will occur later into the year. Both adults and nymphs feed by piercing individual leaf cells and sucking out the contents. Light infestations cause leaves to appear stippled due to the death of individual cells. Heavier feeding results in brown, dried leaves that fall prematurely. Some studies have shown that light infestations can actually improve the quality of grapes on vines that are overly vigorous, but excessive feeding reduces the photosynthetic activity of the vines, resulting in delayed maturity, yield losses, and reduced fruit quality.

Large infestations deplete carbohydrate stores and weaken the vines. Light coloured varieties can become spotted and unsightly with excrement, which is a particular concern for table grapes. Adult leafhoppers are also an annoyance to pickers during harvest. Neither species is known to transmit viruses or other plant diseases.

MONITORING AND SPRAY THRESHOLDS

Overwintered adults usually congregate in higher numbers on the edges of vineyards or in sheltered locations and these areas should receive particular attention in the spring. Leafhoppers also prefer vigorous vines and those that leaf out earliest. Monitoring numbers of adults captured in spring on yellow sticky cards can help determine relative infestation levels, but because some eggs fail to develop and predation rates are variable, these numbers do not correlate well with subsequent nymph counts. It is helpful to note where large numbers of adults occur early in the season, as these areas should be observed more closely later on.

Adults are more tolerant of insecticides and sprays should be targeted against the wingless immature stages. Depending on the location of the vineyard, monitoring of 1st generation nymphs should begin in early to mid June. Monitor populations by counting numbers of nymphs on the lower surfaces of 5-10 leaves in at least 5 locations per block or variety. Select older bottom leaves up to just above the fruiting zone. In order to prevent damage to any vines, sampling should focus on more heavily infested areas. Sample from the north and east sides of rows where more nymphs will be found and focus on the leaf zone showing signs of feeding damage. Infestation levels are based on the average number of nymphs per leaf.

Established grapevines are tolerant of leafhopper damage during the first half of the season when growth is rapid. Because of this, there are no established thresholds for 1st generation nymphs. The threshold of approximately 20-25 nymphs per leaf established for the 2nd generation in late summer can also be used as an approximate threshold for early summer sprays. A higher threshold can be used for 1st generation nymphs if healthy populations of predators are present and a high rate of parasitism is likely. Control of 1st generation nymphs is often more effective as there is less foliage, nymphs are confined to leaves around the fruiting zone, and sprays can be directed in a narrow band to ensure good coverage. It is recommended, therefore, to treat vineyards with a history of heavy leafhopper damage during this time. Soap sprays are much more effective when targeted against 1st generation nymphs when approximately 80% of the eggs have hatched. Egg hatch can only be determined with the aid of a low power microscope, but the presence of mostly small, young nymphs can be used as an approximate measure. Because soaps are not as effective as other insecticides, the threshold for vineyards that are managed organically should be reduced considerably, perhaps to around 5-10 nymphs per leaf.

Second generation nymphs should be monitored beginning in August in a manner similar to that outlined above. Greater attention should be paid to vines in the center of the field, however, as infestations will have spread from field edges and 'hot spots' to a larger area of the vineyard. Nymph counts should be collected initially every other week, and then weekly when the time for spraying approaches.

For wine grapes, an approximate threshold of 20-25 nymphs/leaf on the more heavily infested leaves can be used as a rough guideline. This is an average value based on counts from several leaves as outlined above. Again, for a particular area of the vineyard, collect at least 5 leaves from 5 areas and calculate the average number of nymphs per leaf. For 2nd generation nymphs, sample leaves from the middle of the canopy above the fruiting zone and from the north or east sides of vines where numbers will be higher. Leafhoppers will be distributed unevenly in a vineyard and it is important to sample thoroughly to determine which areas require treatment. Although the above threshold provides a rough guideline, an acceptable spray threshold should be determined by individual growers based on their past experience and additional factors such as vine vigour and numbers of beneficial insects.

BIOLOGICAL CONTROL

Birds, spiders, predators, parasites, and diseases attack leafhopper eggs, nymphs or adults. A small egg parasite, *Anagrus daanei*, helps control VCLH in some vineyards where parasitism rates approach 100% late in the season. A different parasite, *A. erythroneurae*, parasitizes eggs of the WGLH and is largely responsible for the control of this pest. *A. erythroneurae* overwinters in eggs of the rose leafhopper on wild and domestic roses, apple, blackberry and plum. In early spring this parasite utilizes eggs of a leafhopper found on a wide range of mints, including catnip and catmint. Most vineyards have large numbers of *A. erythroneura* and outbreaks of WGLH are largely associated with sprays of insecticides that do not control this leafhopper but are toxic to the parasite. Excessive rates or unnecessary sprays of insecticides will also reduce numbers of other beneficial insects and spiders.

CULTURAL AND PHYSICAL CONTROL

Leafhoppers prefer excessively vigorous plants and vines should be irrigated and fertilized to maintain moderate growth best suited to the production of high quality wine grapes. The use of yellow sticky tape applied below the cordon in spring can be an economical way to manage leafhoppers in some vineyards. Although costly, this method helps preserve beneficial insects and reduces or eliminates the need for additional insecticides later in the season. The use of sticky tape is likely not cost effective for large areas but it can be a useful method where infestations are confined to small areas or to the edges of vineyards.

A recently completed study demonstrated that removal of basal leaves early in the season (mid to late June) when most eggs of the 1st generation have been laid effectively reduced numbers of leafhoppers and the incidence of bunch rots. No significant differences were noted in fruit quality other than a slight reduction in yield and berry size, but it is recommended that growers first assess the effectiveness of early leaf removal on only a few vines of each variety. Removal of leaves around the fruiting zone at any time of the season might not be suitable for stressed vines or on sites with intense heat and light that might lead to sunburn. This practice is labour intensive, but growers often remove these leaves later in the season in order to improve grape quality. Removing basal leaves too early will force leaf-hoppers to deposit their eggs higher up the vine and might inhibit bud initiation required for the following year.

CHEMICAL CONTROL

Some insecticides registered for the control of leafhoppers are ineffective for the WGLH and it is important to determine if this species is present in your vineyard. If only the VCLH is present, natural enemies should be preserved by applying the lowest effective rate only to areas where the pest has exceeded the threshold level. To date, this species has not yet developed resistance to insecticides.

When monitoring of leafhopper nymphs indicates that insecticide sprays are required, apply one of the materials listed in*Pest Control Products Recommended for Use on Grapes in British Columbia*.

It is often sufficient to spray small areas or treat only the perimeter of the vineyard. Spraying in this fashion is more economical and helps preserve beneficial insects. Although reduced rates (640 ml/ha) of carbaryl (Sevin XLR Plus™) suggested in earlier editions of this guide to help preserve beneficial insects are likely sufficient to control VCLH, full label rates will provide only moderate control of the more resistant WGLH. Malathion is currently registered for the control of leafhoppers on grapevines but it is unlikely to be effective against the WGLH. Assail™ (acetamiprid) has been shown to be repellent to adults and will cause them to move to unsprayed portions of the vineyard. Sprays should be targeted against small nymphs. Although Assail is less damaging to some beneficial insects compared with many broad-spectrum insecticides, it is toxic to predacious thrips and can cause spider mite numbers to increase. Safer's Insecticidal Soap applied for the control of mites and grape mealybug will reduce leafhopper numbers somewhat. Apply soap in high volume sprays and ensure that the bottom leaf surfaces are covered thoroughly. Recent research at AAFC-PARC Summerland has shown that some newer fungicides and surfactants are repellent to adult leafhoppers and their use might eliminate the need for additional sprays. Increased movement of adult leafhoppers within and between vineyards requires greater vigilance, however.

CLIMBING CUTWORMS (*LEPIDOPTERA: NOCTUIDAE*)

To date, 17 species of cutworm have been identified feeding on the buds of grapevines in the southern interior of British Columbia; the three most common species being *Abagrotis orbis* (Grote), *A. reedi* Buckett, and *A. nefascia* (Smith). Damage can vary greatly between years and locations and it is not uncommon for only parts of vineyards to be seriously damaged. Because populations fluctuate greatly and monitoring of cutworm larvae is difficult, damage can be considerable.

IDENTIFICATION

Cutworm larvae are smooth, soft, thick-bodied caterpillars measuring about 4 to 5 cm when fully grown. Larvae typically live in the soil or under debris and will curl up when disturbed or handled. Depending on the species, larvae are most often light to dark brown, and are usually marked with spots or stripes. Adults are drab moths with bodies around 2 to 2.5 cm long. The hind wings, which are sometimes distinctly marked, are usually lighter in colour than the forewings. Adults are active at night and rest during the day in sheltered locations under bark, debris, or leaves.

Cutworm larvaPhoto courtesy of M. Weis, Agriculture and Agri-Food Canada, Summerland.

LIFE CYCLE AND DAMAGE

Cutworms that feed on grapes spend the winter as small larvae in the soil or under fallen leaves or debris. When the weather warms in spring they resume feeding and complete their development. Damage to grapes occurs during this time when the rapidly growing larvae feed on unopened buds and new shoots. Moths emerge in June and remain in a quiescent state throughout the summer until fall when they become sexually mature and begin laying eggs that hatch into the overwintering larvae. There is a single generation each year.

Damage to grape buds varies with the size of the cutworm larvae. Small larvae will often burrow into a bud, creating nearly circular holes that were previously attributed to feeding by adult click beetles. Larger cutworm larvae will chew large, somewhat ragged holes and remove a large portion or all of the bud. Feeding on new, tender shoots can cause the shoots to break or wilt. Larvae climb the vine trunks or support posts each night and return to the soil during the day, or they will find refuge under loose bark of older vines.

Cutworm damage to grape bud Photo courtesy of H. Buchler. Cutworm damage to young shoot.Photo courtesy of M. Weis, Agriculture and Agri-Food Canada, Summerland.

MONITORING AND SPRAY THRESHOLDS

Cutworm numbers vary over time and between and often within vineyards. Some vineyards rarely experience significant levels of cutworm feeding damage while others are damaged yearly. Mild, dry conditions during the winter increase survival rates, while outbreaks of viral and other diseases that affect cutworms can reduce numbers to low levels. Cutworm damage is often higher in areas with dry gravelly or sandy soils that do not support good stands of broadleaf weeds, with drip irrigation systems, or as a result of weeds being controlled too early in the spring before shoots have elongated. Maintaining a record of previous damage can help suggest which areas need to be monitored most closely and where damage is likely to occur first, but bud damage needs to be assessed throughout the vineyard. New vines planted in untreated land previously in pasture are prone to damage, and larvae are often sheltered and somewhat confined within the protective milk cartons or sleeves placed around new vines. Loss of buds on new vines can retard growth and delay fruiting and they should be closely inspected in spring for signs of damage and treated when necessary.

Significant amounts of damage can occur over a period of two or three days when the nights remain warm, and it is especially important to monitor for damaged buds frequently during this time. At other times it is usually sufficient to monitor vines weekly or twice weekly until shoots have elongated. An accurate threshold based on the degree of bud damage is difficult to determine, as it is influenced by the variety, training system and method of pruning, presence of predators and parasites, and tolerance level of the individual vineyard manager.

Suggested economic thresholds from growers range from virtually zero to over 15%; a value around 3% bud loss is likely for most *vinifera* varieties. Establishment of spray thresholds should also consider the potential for secondary pest outbreaks that will require additional insecticide sprays later in the season.

BIOLOGICAL CONTROL

Cutworm eggs are consumed by various predators and are parasitized by small wasps. Larvae are eaten by ground beetles, hunting spiders and birds, and are parasitized by wasps and flies. Fungal and bacterial diseases that attack cutworm larvae are very common, particularly when conditions are cool and wet. The combined effect of these natural controls is to eliminate most cutworm eggs and larvae. Of the 300 or more eggs laid by a female moth, only a few survive to become reproductive adults. Even so, damage resulting from the few remaining larvae can be considerable.

CULTURAL CONTROL

Broadleaf weeds should not be controlled in spring until shoots have elongated and the first leaves have expanded. Damage is generally much lower in vineyards with mixed ground cover vegetation and where broadleaf weeds, particularly winter annual mustards such as *Draba verna* and Shepherd's purse, *Capsella bursa-pastoris*, are present in the vine rows in spring. Larvae are usually more common in areas with well-drained sandy or sandy-loam soils that support fewer broadleaf weeds.

Diseases that infect larvae are also likely to be more prevalent in soils that retain moisture. Control of vegetation with herbicides or repeated discing during the fall prior to the establishment of a new vineyard will help reduce damage to newly-planted vines.

Cutworm damage is partially related to pruning practices and grape varieties. Vines with a greater number of retained buds in spring will suffer marginally less damage, but there will be added costs related to removal of extra shoots. Some grape varieties will produce fruit on secondary shoots, but yields will still be reduced and ripening can be delayed or uneven.

CHEMICAL CONTROL

Sprays of the pyrethroid insecticide Pounce (permethrin) should be directed to the soil at the base of vines, to vine trunks, and to support posts so that larvae contact the insecticide while moving from the soil to the canopy. Do not cultivate or disturb the soil for several days after treatment. For young vines, sprays should be directed into the milk cartons or other protective structures applied around newly planted vines. Avoid spraying insecticides unnecessarily, as these are very damaging to beneficial insects and can result in outbreaks of secondary pests later in the season.

GRAPE PHYLLOXERA (*DAKTULOSPHAIRA VITIFOLIAE* (FITCH)

Grape phylloxera (Homoptera: Phylloxeridae), native to eastern North America, are a type of aphid found on the roots of grapes or in galls formed on the leaves of wild grapes or hybrid varieties. European *vinifera* varieties are particularly susceptible to this pest, and the accidental introduction of phylloxera to California, Europe, and elsewhere initially devastated these industries. Management of phylloxera with insecticides was not successful and the wine grape industry was only rescued with the development of resistant rootstocks derived from American *Vitis* species or hybrids between these and susceptible *V. vinifera*. Phylloxera were first identified in the Penticton area in 1961; infestations currently are scattered and at relatively low levels in British Columbia vineyards, mostly due to the use of resistant rootstocks and the failure of this pest to thrive on sandy soils.

As early as 1914 it was recognized that grape phylloxera consisted of several distinct biotypes or races. Biotypes differ in their ability to form galls on the leaves of certain cultivars, in the degree of feeding damage caused to roots, and their propensity to form sexual or winged forms. One study demonstrated that rootstocks with unsatisfactory levels of resistance to phylloxera in Europe and South Africa were suitable for California, Australia and New Zealand. Loss of resistance can also occur over time as new biotypes of phylloxera arise.

IDENTIFICATION

Adults are very small, usually wingless, and less than 1mm long and half as wide. Colour varies from pale green, yellowish-green or light brown on vigorous roots, to brown or orange-brown on declining roots; as adults age they become a darker brown or even purplish-brown. Winged sexual forms are rare in British Columbia and the northwestern U.S. and they do not appear able to reproduce sexually on *vinifera* grapes. Leaf galls are formed on French-American hybrids, Concord juice grapes and some rootstocks. The relatively small galls on the undersides of leaves are green and roughly circular. These should not be confused with the white or brown fuzzy galls of variable sizes and shapes formed on the undersides of leaves by erineum (leaf blister) mite.

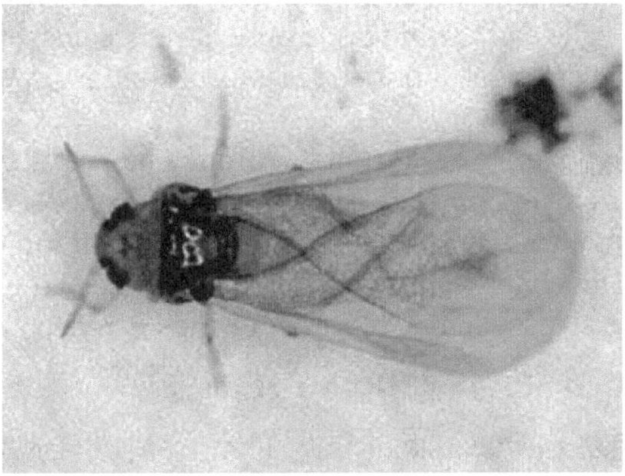

Grape phylloxera adults and eggs on grape roots. Photo courtesty Agriculture and Agri-Food Canada Grape phylloxera - winged adult

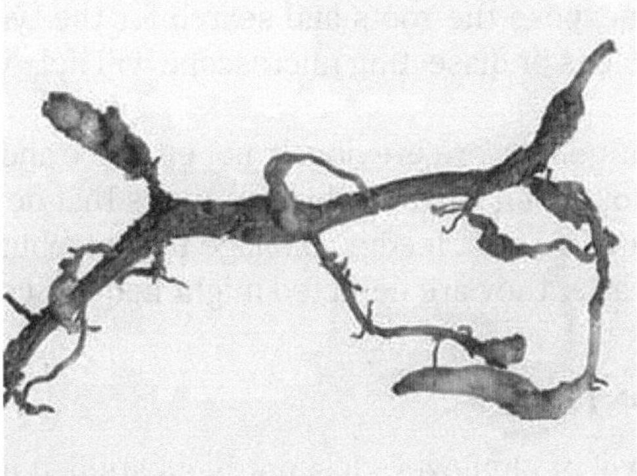

Phylloxera-induced galls on grape roots.Photo courtesty Agriculture and Agri-Food Canada Phylloxera galls on leaf of grape hybrid

LIFE CYCLE AND DAMAGE

Phylloxera overwinter on roots as small, dark hibernating nymphs that complete development in spring to become females that reproduce without having mated. Two or three generations are produced each year. When populations are large or the vines are declining, some newly hatched nymphs leave the roots through cracks in the soil and migrate to the roots of nearby vines. These small crawlers have been known to climb vine trunks and be blown on the wind for some distance. In late summer a portion of the nymphs develop into winged females (alates) that leave the soil and fly to other vines where they lay eggs in the crevices in the bark that hatch into males and females. After mating females produce a single larger overwintering egg. In spring the eggs hatch into females that feed on developing leaves and produce galls on susceptible varieties. Nymphs produced in the galls can infest other leaves or migrate down to the roots. Most grapes grown in British Columbia are not susceptible to leaf galls and the life cycle here largely consists of asexual forms confined to the roots of grapes. For this reason, infestations spread at a relatively slow rate.

Feeding of phylloxera on the roots of grapevines results in severe water and nutrient stress and eventual death of the vines. Roots become stunted and galled and die prematurely from the injection of harmful saliva and invasion of fungal pathogens. Depending on vine vigour and the size of the infestation, vines can slowly become stressed and chlorotic, or they can decline rapidly.

MONITORING AND SPRAY THRESHOLDS

Vines that fail to thrive and become chlorotic and unproductive should be inspected for the presence of phylloxera. Galls on leaves are an obvious indication of phylloxera on hybrid varieties, but damage confined to roots of *vinifera* wine grapes can be confused with damage from nematodes or root diseases. Carefully expose the roots and search for the typical swellings and dying roots. A hand lens or dissecting microscope will help verify the presence of phylloxera.

Insecticide treatments were previously not effective and there are no spray thresholds for phylloxera on roots. Hybrid cultivars that develop leaf galls are better able to tolerate or resist feeding damage to roots, but elimination of the leaf gall forms soon after they are detected might help reduce the extent of the infestation.

BIOLOGICAL CONTROL

Biological control of phylloxera has not been studied in detail. A number of generalist predators are known to feed on phylloxera in leaf galls. In Ontario, larvae of two species of predaceous flies were commonly associated with leaf galls. Predaceous soil-dwelling carabid beetles undoubtedly feed on phylloxera infesting roots, and because they do not produce honeydew, ants are likely to feed on phylloxera rather than protect and 'milk' them. Although natural enemies are unable to effectively control populations of phylloxera, diseases and predators will help keep populations low and reduce the rate of spread.

CULTURAL CONTROL

When establishing a new vineyard, purchase plants from reputable suppliers and ensure that planting material is healthy and free of phylloxera.

Note that grape plants or plant parts for propagation moving from phylloxera-infested areas of Canada into British Columbia, must be accompanied by a Movement Certificate issued by the Canadian Food Inspection Agency (CFIA), and must be treated with a CFIA approved treatment prior to movement. If you wish to source grapevine material from outside of B.C., please contact your local CFIA office to get details on the requirements.

To ensure that phylloxera are not moved from infested to clean areas on farm machinery, disinfest all equipment. Less frequent cultivation that penetrates the root zone is thought to be helpful.

Resistant rootstocks provide an effective means of combating phylloxera. Almost all resistant rootstocks were developed from crosses between several species of wild grapes native to eastern and southern North America (*V. riparia*,*V. rupestris*, *V. berlandieri* and others). Some of the more common grape rootstocks with good phylloxera resistance include SO4, 5BB, 5C, 420A, and 161-49C. Rootstocks vary in characteristics such as vigour and it is often difficult to select an appropriate rootstock for a particular vineyard. Soil types, resistance to nematodes, vine spacing and several other factors need to be considered in addition to phylloxera resistance. Prior to planting it is advisable to read appropriate publications and consult with vineyard managers or consultants.

Vigorous vines resist phylloxera damage better than stressed vines. Provision of additional water and nutrients can help delay the rate of decline. Infested vines should not be heavily cropped and extra care is required to prevent additional stresses from insects and diseases.

Some varieties are more vigorous and are better able to resist damage. Phylloxera infestations are more damaging and spread more rapidly on heavier soils that are prone to cracking, while vines planted in sandy soils can be nearly immune to attack.

CHEMICAL CONTROL

Movento™ (spirotetramat), a systemic insecticide effective against plant-feeding insects with sucking mouthparts, was recently registered for the control of grape phylloxera and mealybug. Sprays should be applied in a minimum spray volume of 500 L/ha as per the label instructions.

WASPS

A number of species of yellowjacket wasps (Hymenoptera: Vespidae) are pests of grape; the most numerous and troublesome being *Vespa pennsylvanica* (Saussure). Two uncommon species of paper wasp (*Polistes* sp.) are native to our area. The European paper wasp, *P. dominulus*, was found in British Columbia in 2003 and has since become very common, particularly around buildings. Yellowjackets, hornets and paper wasps are important predators of grape pests throughout most of the year, but in fall they often feed in large numbers on damaged fruit. Wasps are also an annoyance to vineyard workers and to pickers during harvest operations. Paper wasps are less aggressive than yellowjackets or hornets and are not as prone to sting.

IDENTIFICATION

Yellowjacket workers are about 1.5 cm in length; the abdomen appears largely yellow with thin black bands. Hornets are slightly larger and appear largely black with yellow or white bands. Paper wasps have thinner bodies with a more pronounced constriction (stalk) between the thorax and abdomen. The body is dark with some yellow markings, but they appear reddish-brown due to the colour of the wings and distal parts of the legs. In flight the legs of paper wasps dangle visibly below the body. All of these wasps produce paper nests, but the brood cells of paper wasps are exposed.

European paper wasp European paper wasp nest in an ornamental shrub

Wasp feeding on grapes. Photo courtesy John Vielvoye, Kelowna B.C.
Wasp damage to grapes. Photo courtesy John Vielvoye, Kelowna B.C.

LIFE CYCLE AND DAMAGE

Yellowjackets and paper wasps overwinter as fertilized females in sheltered locations. These queens seek suitable nest sites in spring and establish a new colony. Nests are initially small, consisting of only a few paper cells, but as workers develop they assist in building up colony numbers and the size of the nest. Depending partly on the species, nests can be in the ground, under the eaves of houses, in hollow trees or the walls of buildings, or exposed in trees. In fall, in addition to workers, colonies produce both males (drones) and queens. When freezing weather arrives, all the drones and workers, as well as the undeveloped larvae and pupae are killed.

Wasps prey on other insects, scavenge food and feed on nectar and the juices of fruit. It has been reported that wasps do not damage fruit themselves but take advantage of injury caused by other insects or birds, diseases such as sour rot, or splitting of fruit. There are contrary reports, however, particularly relating to the European paper wasp, that they are able to directly feed on intact fruit. As mentioned above, the greatest detriment is from annoyance or injury to pickers. Their stings result in swelling and a painful itch and can trigger severe allergic reactions in some individuals.

MONITORING AND SPRAY THRESHOLDS

Wasps are generally considered to be beneficial or of little economic importance. There are no established thresholds and the need to spray is determined by the tolerance level of individual vineyard managers. The recent arrival of the European paper wasp will likely make it necessary to control wasps more frequently.

BIOLOGICAL CONTROL

Wasp colonies can be attacked by other wasps. Eastern and western kingbirds feed on foraging adults and larvae are subject to certain diseases. The onset of winter eliminates established colonies and overwintering queens often succumb to cold or are eaten by other insects, or even by mice.

CULTURAL CONTROL

Elimination of nests early in the season can be an effective management strategy, but they are often located outside the vineyard or well hidden. Aerial nests in trees or under eaves can be removed and placed in a plastic bag at night when wasps are inactive. Nests in walls or natural cavities can be treated by blowing or puffing commercial wasp control products into the entrance holes at night. Sealing of entrances can help prevent future nests.

Trapping can be used to reduce numbers of foraging wasps in localized areas. Commercial wasp traps are available, or homemade versions can be made from recycled 1 litre plastic pop bottles. Sugary solutions are contained in the bottom of the trap and wasps that enter the small holes eventually fall into the liquid solution and drown.

Trapping early in the season when populations are low is generally more effective than attempting to remove large numbers of wasps later in the season. For additional information on social wasps, management strategies, and design of wasp traps, visit the University of California Statewide Integrated Pest Management Program.

CHEMICAL CONTROL

There are currently no insecticides registered specifically for the control of wasps feeding on grapes.

SPOTTED WING DROSOPHILA (FRUIT FLY) PEST ALERT

Spotted wing drosophila (*Drosophila suzukii*), a serious new fruit fly pest of soft fruit and berries, was first identified in British Columbia in 2009. It is now widespread in Coastal and Interior fruit growing areas of B.C.

Spotted wing drosophila is a temperate fruit fly, native to Southeast Asia; preferring temperatures of 20-30 °C. It is known to infest thin-skinned fruit. Many species of fruit flies are present in late summer; most normally infest overripe, fallen, decaying fruit, so are not crop-limiting pests. However, a spotted wing drosophila female lays her eggs inside sound fruit before harvest with her saw-like ovipositor, which contaminates fruit with larvae, and causes it to become soft and unmarketable.

APPEARANCE

Many features are typical for *Drosophila* fruit flies, with a few key differences. Male and female characteristics are key identifiers for this species.

Adults: 2-3 mm (1/8 inch) long, brownish with red eyes and clear fly-like wings. Compared to other fruit flies, *D. suzukii* is a robust fly, but this is difficult to discern unless compared directly to other species. Males have a black/grey spot on the end of each wing, as well as two black 'combs' or bands on the front legs. The females do not have spots or leg bands. Females have saw-like ovipositors that are used to cut into fruit skin. Ovipositors are easier to see when extended. A hand-lens or dissecting microscope is needed to confirm ovipositor presence.

Eggs: 0.6 mm long, oval, white, 2 filaments at one end.

Larvae: Legless, headless, up to 6 mm long at maturity, white or transparent.

Pupa: 3 mm long, brown, football-shaped, two stalks with small finger-like projections on one end. Refer to the Identification Guide for Spotted Wing

Drosophila (PDF, 2.5 MB) for additional information on characteristics of this pest.

DISTRIBUTION

SWD continues to spread, and is now widely distributed globally in most temperate soft fruit producing areas. Notable exceptions are South America, New Zealand and Australia, where SWD is not known to be established. Below is an approximate list of places where SWD is established.

Asia: Burma, China, India, Japan, Korea, Russia (far east), Thailand, Taiwan, Pakistan, Myanmar

North America: Canada, United States, Mexico

Europe: Spain, Russia, France, Italy, Portugal, Germany, Slovenia, Switzerland, Croatia, Belgium, Austria, United Kingdom, Netherlands

HOSTS

In British Columbia, spotted wing drosophila has been confirmed infesting wild and cultivated raspberry and blackberry (*Rubus*), blueberry (*Vaccinium*), strawberry (*Fragaria*), cherry, peach, nectarine, apricot, plum (*Prunus*), and suspected in hardy kiwifruit (*Actinidia)*. Wild hosts confirmed in Coastal B.C. include Oregon grape (*Mahonia aquafolium*), elderberry (*Sambucus*), currant (*Ribes*), dogwood (*Cornus kousa*), mulberry (*Morus*), salmonberry (*Rubus spectabilis*), thimbleberry (*Rubus parviflorus*), salal (*Gaultheria shallon*), Indian plum (*Oemleria cerasiformis*) wild *Prunus* species, and red huckleberry (*Vaccinium parvifolium*). In Interior B.C, wild hosts confirmed include Oregon grape (*Mahonia* sp.), blue elderberry (*Sambucus cerulean*), Northern black currant (*Ribes hudsonianum*), Tatarian honeysuckle (*Lonicera tatarica*), Mahaleb cherry (*Prunus mahaleb*), and golden currant (*Ribes aureum*) (H. Thistlewood, AAFC, Summerland).

DAMAGE

Females lay eggs under the skin of ripe fruit shortly before harvest. Larvae hatch and begin to feed within the fruit, causing softening in the area of feeding. There can be several larvae in a fruit, which hastens softening and fruit collapse. Holes the size of pin pricks are evident within the soft areas of infested fruit. These holes result from egg laying and are used as breathing holes by larvae. In addition, these holes provide entry points for diseases such as brown rot and botrytis.

LIFE CYCLE

Spotted wing drosophila emerging in the fall overwinter as adult flies. In spring flies become active, mate and lay eggs in ripening fruit. Based on climate model predictions, there could be up to 5 generations per year in B.C.

Generations will likely be overlapping as flies are relatively long-lived particularly at temperatures of 20°C and cooler. Based on a Japanese publication (Kanzawa 1939), oviposition lasts 10-59 days, with 7-16 eggs laid per day, and averaging 384 eggs per female. Eggs hatch in 2-72 hours, larvae mature in 3-13 days, and pupae reside in fruit or outside of fruit for 3-15 days. In the lab at constant temperature, one generation takes 50 days at 12°C, 21-25 days at 15°C, 19 days at 18°C, 8.5 days at 25°C, and 7 days at 28°C. Adults are also attracted to dropped and decaying fruit and will feed on it. SWD will complete its development in dropped fruit.

How it Spreads

Spotted wing drosophila adults can be blown by wind to nearby locations. However, long distance dispersal is through transportation of infested fruit to new regions. Non-fruit bearing plants are not considered to be of significant risk to transport this pest.

MANAGEMENT

Management recommendations include good harvest and sanitation practices, such as culling soft fruit, destroying culls, and keeping processing areas and equipment free of old fruit. Please refer to the factsheets on management linked below under "further information" for more details on managing this pest in berry and tree fruit crops. Emergency registration products approved for the control of spotted wing drosophila in B.C. for the 2013 season are Delegate (spinetoram), Entrust (spinosad), Malathion (malathion), Pyganic (pyrethrins) and Ripcord (cypermethrin). Both Entrust and Pyganic are acceptable for organic crop production. Labels are posted at: Current Emergency Registrations for British Columbia.

Quarantine Regulations

This pest is not regulated in the United States and Canada. The regulatory status of this fly in other countries should be checked with packers.

Surveillance

Area-wide surveillance with apple cider vinegar traps in British Columbia for SWD indicates that flies are present and active throughout the year in the Fraser Valley, though numbers are very low in February through May. In the fruit growing interior regions, SWD can be caught in traps from May until November. Surveillance continues in fruit growing regions of B.C. in 2013 to determine when SWD flies are active in commercial fields.

THRIPS

Several species of thrips can be found feeding, at least occasionally, on grapes; the three most common are the western flower thrips (WFT), the grape thrips (GT), and the flower thrips (FT). Thrips are generally minor pests of wine grapes, but are important pests of table grapes due to scarring of fruit. Their populations often increase following applications of broad-spectrum insecticides that reduce numbers of beneficial insects. All species feed in a similar manner, but differences in biology and timing of infestations results in differential damage to grapes.

IDENTIFICATION

All species have long, slender bodies with dark protruding eyes and relatively short beaded antennae. Adults have two pairs of wings that are heavily fringed along the edges. WFT are dark yellow to brown and 1- 1.5 mm long. GT and FT are both around 1 mm long, the former is amber yellow with a yellowy-orange thorax, and the latter is uniformly pale yellow to nearly white. Nymphs of all species are similar in appearance to the adults except that they are smaller, slightly paler in colour, and lack wings. Thrips tend to be very active; adults not only fly but also run and hop rapidly.

Plant feeding thrips should not be confused with predaceous thrips. Adults of beneficial thrips may be dark black or purplish black with clear wings or wings that appear light grey with white bands, or they are pale with dark bands on the wings. They are slightly larger to nearly twice as large as the pest species. Nymphs of predaceous thrips have bodies that are banded yellow and orange, are reddish purple in colour, or are pale with the joints of the appendages dark. Compared with phytophagous species, predaceous thrips have front legs that are enlarged and slightly curved that they use for grasping prey. These features and the slightly larger mouthparts are best observed under a dissecting microscope.

LIFE CYCLE AND DAMAGE

Adult female thrips overwinter under leaves and plant debris on the ground. In spring they deposit their eggs into developing leaf, stem, and flower tissues. The time from egg to adult is about three weeks during warm weather, and several generations are produced yearly. Most thrips feed on a wide range of host plants, and large numbers will often move to a succession of new succulent hosts throughout the year. Both adults and nymphs feed by rasping the leaf surface and sucking up the escaping fluids.

WFT adults enter vineyards in early spring, often in large numbers, when saskatoon bushes have finished flowering and other native vegetation is beginning to dry down. In addition to ground cover plants, WFT feed on flower and fruit tissues of grapes from bloom to just after fruit set. Their feeding under caps and persistent flower parts causes berry scarring and russeting. White and green grape varieties are also damaged when females insert their eggs into fruit, causing dark spots or halo spots. Feeding on developing shoots is occasionally so heavy that shoots are stunted and leaves deformed, but damage of this sort often only occurs around the perimeter of vineyards. WFT usually decline greatly in numbers following fruit set. FT have a similar biology and contribute to the damage caused to fruit in spring, but they are not as numerous as WFT. GT are generally only found in significant numbers on grape during the summer months, and their feeding is largely confined to leaves and shoots. Numbers can build up rapidly, however, particularly following insecticide treatments for other grape pests. Leaf veins can be scarred and fail to expand properly, leaf edges can become necrotic, and shoot growth inhibited, resulting in shortened internodes.

MONITORING AND SPRAY THRESHOLDS

WFT often invade vineyards in large numbers for only brief periods in spring and it and the FT are not normally an economic concern for wine grapes. Scarring of fruit has not been shown to affect wine quality and damage to leaves and shoots is usually minor.

Damage usually occurs over a short period of time, which makes early detection and control difficult. Inspect table grapes for the presence of WFT and FT in spring from around 75% bloom to the end of fruit set and apply insecticides as required. Verify the presence of thrips by sharply striking fruit clusters or leaves against a dull white or green coloured surface. The decision to spray should be based partly on the table grape variety, records of past damage, and the intended market for the fruit. There are no firm thresholds for thrips infesting grapes in British Columbia.

Watch for GT damage to young leaves and shoots during summer and early fall. Use an eyepiece or magnifying glass to verify that thrips are present. As for fruit clusters, leaves and shoots can be held over a piece of white cardboard and sharply hit to dislodge the thrips. Control on mature vines is usually only warranted when growth is stunted sufficiently early in the summer to delay ripening and affect growth the following season.

BIOLOGICAL CONTROL

Predatory bugs such as the minute pirate bug (*Orius tristicolor)*, anthocorids (*Anthocorus melanocerus*), and Deraeocoris (*Deraeocoris brevis*) are all known to feed on thrips. Studies in the Okanagan and Similkameen Valleys have indicated that predatory thrips are important regulators of pest thrips populations. In the laboratory, predatory thrips at a density of one every 10 to 15 leaves is sufficient to control WFT.

CULTURAL CONTROL

So that thrips are not forced to feed on table grapes when they are most susceptible to fruit damage, avoid mowing cover crops during bloom and fruit set.

CHEMICAL CONTROL

Success™ (spinosad) was recently registered for suppression of thrips on grapes. Insecticidal soap applied for mites and grape mealybug will also help suppress thrips populations.

SPIDER MITES

The main species of spider mites infesting grapes in British Columbia are the European red mite (ERM) and the two-spotted spider mite (TSSM). They differ somewhat in biology, but feeding damage and management strategies are largely the same. Spider mites should be considered secondary or induced pests of grapes; their numbers increase following the use of broad-spectrum insecticides that reduce numbers of spider mite predators. Although it is somewhat selective in action and preserves populations of some predators, the insecticide Assail is toxic to predacious thrips that are effective predators of mites in B.C. vineyards. Studies in Washington State have also shown that the active ingredient in Assail and related materials increases spider mite reproduction. Due to its persistence, increases in mite numbers can occur up to several months after application.

IDENTIFICATION

Adults of both species are small, around 0.5 mm in length, and just barely visible to the naked eye. They are nearly spherical in shape and have eight legs and short piercing mouthparts. The ERM is dark red in colour with long pale hairs (setae) projecting from small pale circular areas on the dorsal surface. The oval eggs are bright orange with a distinctive hairlike projection on the upper side. The TSSM varies in colour from pale yellow to greenish yellow to bright orange; the common pale form has a distinct dark dorsal spot on each side of the body. Its eggs are white. Both species feed on the undersides of leaves. TSSM usually congregate in clusters and produces large quantities of silk webbing, while ERM produce less webbing and are more evenly distributed over the leaf surface.

LIFE CYCLE AND DAMAGE

ERM, our most common species, pass the winter as eggs laid on vine canes and trunks. Eggs hatch in spring and the young microscopic nymphs begin to feed on young foliage. TSSM overwinter as fertilized females under the bark or in other sheltered areas around the base of vines. When warmer weather arrives in spring the females begin to feed and deposit eggs. Both species can produce around six to eight overlapping generations each season, and all stages can be found on grapes at any time during the summer months. Females of both species are capable of producing 200 or more eggs each and development can be rapid during warm weather. Spider mite populations can, therefore, explode rapidly under favourable conditions. Both adults and nymphs feed by piercing individual leaf cells and removing the fluid contents. Healthy grapevines can tolerate moderate numbers of spider mites, which will cause chlorotic spots on the leaves.

Heavy feeding results in brown leaves that fall prematurely; reducing photosynthetic activity and vine vigour. Heavy feeding damage can delay ripening of fruit. The large amount of webbing produced particularly by the TSSM is also a cosmetic problem for table grapes.

MONITORING AND SPRAY THRESHOLDS

Spider mites rarely cause significant damage to grapes prior to mid-summer. No firm thresholds have been developed for spider mites on grapes in British Columbia, but approximate guidelines can be suggested from research conducted in other countries.

In Europe, a threshold for ERM on grapes was determined to be 60 to 70% infested leaves in spring, while in summer leaves with 1 or more mites should not exceed 30 to 45% of total leaves on a shoot. Because a similar threshold has been suggested for TSSM in Switzerland, numbers of both species can be combined into a single spider mite count. A good hand lens or low-power dissecting microscope would help provide accurate counts of infestation levels. A fairly reliable method is to monitor damage to leaves and spray when a moderate amount of 'bronzing' has occurred, around 10% defoliation. Because mite numbers can increase very rapidly it is necessary to monitor grapes frequently during hot weather.

When monitoring for mite damage it is useful to consider previous infestation levels. Vineyards with low chemical inputs that have not experienced severe spider mite outbreaks in the past are unlikely to require treatment, while those that have been treated with broad-spectrum insecticides should be observed closely.

BIOLOGICAL CONTROL

Spider mites were of minor importance prior to the widespread use of synthetic pesticides. Due largely to the harmful effects of pesticides on populations of predators, spider mites are now the most important pests of grapes in many regions of the world.

The first course of action for the management of spider mites is the preservation and enhancement of beneficial species. Pesticides should only be applied when necessary and only to parts of the vineyard where pest populations are sufficiently high to warrant control. Whenever possible select materials that are least damaging to non-target species. Numbers of spiders and beneficial insects are usually higher in vineyards with permanent, mixed groundcover that provides pollen and alternate sources of prey. Proximity to uncultivated or unsprayed areas allows beneficial insects to re-colonize vineyards following spray treatments. Several species of predatory mites (e.g. *Metaseiulus*, *Typhlodromus* and *Amblyseius* species) feed on spider mites and their eggs. Most beneficial mites are a pale opaque colour but they can appear light orange if they have been feeding on ERM. They are oblong in shape and slightly smaller than adult spider mites. The long-legged orange whirligig mite (*Anystis agilis*) is uncommon on grapes in British Columbia Like most other predatory mites, this large species is more active than its prey. Predatory mites are very sensitive to a range of pesticides, including sulphur fungicides and pyrethroids such as permethrin (Pounce™).

A number of spiders and predatory insects feed on spider mites or their eggs. Many species, such as the minute pirate bug (*Orius tristicolor*), are generalist predators that do not specialize on mites, while others such as the aptly named spider mite destroyer, *Stethorus picipes*, are very effective predators that feed almost exclusively on mites. *Stethorus picipes* is a small, dark species of ladybeetle with a slightly hairy appearance. The elongate bodies of the larvae are also nearly black in colour with a body covered with numerous hairs. At least four species of predatory thrips can be found in Okanagan Valley vineyards; adults of three of these are black, while the fourth, the six-spotted thrips, *Scolothrips sexmaculatus*, is pale with dark spots on the wings. All 4 species feed on spider mites, and healthy numbers of these predators are usually associated with low populations of their prey. The western flower thrips is considered a pest of grapes, but it is also known to feed on the eggs of spider mites.

CULTURAL CONTROL

Vineyards can be managed in ways that help alleviate mite problems. Hot, dry and dusty conditions favour the buildup of spider mite populations. Where mites are a problem, roadways can be treated with oils or other materials to reduce dust. Vineyards that are continuously cultivated will produce more dust than those with permanent ground covers. Moreover, a permanent mixed groundcover will support greater numbers of beneficial insects and predaceous mites. Compared with bare soil, planted aisles will decrease air temperatures and raise humidity levels somewhat, providing less favourable conditions for spider mites. Spider mite populations can become elevated on vines suffering severe drought stress, and these plants are also less able to tolerate damage. Overhead irrigation will help reduce mite infestations.

CHEMICAL CONTROL

A number of miticides (e.g. Acramite™, Kelthane™ , Envidor™ and Pyramite™) are registered on grapes for mite control. Insecticidal soap applied for the control of leafhoppers will help suppress mites.

GRAPE ERINEUM MITE

IDENTIFICATION

Adult erineum mites are only 0.2 mm long and a dissecting microscope is required to observe them properly. Their creamy pale bodies are elongate with only two pairs of legs. Although erineum galls can be confused with certain fungal diseases or phylloxera leaf galls by inexperienced growers, their characteristic appearance is the best way to identify these pests. The hemispherical concave galls or erinea formed on the undersides of leaves initially have a felty white appearance due to the presence of densely packed long leaf hairs. Opposite to these galls, bumps will be visible on the upper surfaces of leaves.

These swellings are a reddish colour in spring and turn green as the leaves mature. Small numbers of mites feed on the dense hairs in each gall, which eventually turn brown. Having been reported recently from Washington State, two other forms of erineum mite, the leaf-curling strain and bud-inhibiting strain are possible in British Columbia. The former causes slight to severe downward curling of leaves in summer with stunting of shoots and increased growth of lateral shoots.

Feeding of the bud-inhibiting strain on buds causes a range of abnormalities, including scarified shoots, short zigzagging internodes, dead terminal buds and malformed leaves. For additional information on erineum mites and other pests of grapes, visit the University of California Integrated Viticulture Online website.

LIFE CYCLE AND DAMAGE

Adult erineum mites overwinter under the scales of buds. Mites move in spring to developing shoots and create galls on young leaves around the fruiting zone. Several generations are produced each year, with new galls developing higher up the shoots. Beginning in late summer, adult mites move back to the buds for winter.

Leaves with a few too many erineum galls appear to function normally and there is little or no economic damage to mature vines until almost all leaves are covered with galls. Even then, damage usually only occurs if vines are also suffering from other stresses.

Establishment of newly planted vines can be delayed somewhat by erineum mite feeding, which will retard or reduce fruiting.

MONITORING AND SPRAY THRESHOLDS

Erineum galls are usually very apparent during routine vineyard operations. Grape varieties differ in susceptibility to this pest and it is important to check individual blocks for the presence of mites. In order to prevent premature loss of leaves, new vines should be inspected and treatments applied if large numbers of galls are found on most leaves during the spring or summer months.

BIOLOGICAL CONTROL

The western predatory mite, *Galendromus occidentalis*, is an important erineum mite predator. Predaceous thrips will feed on exposed mites, but their effectiveness is likely limited when mites are protected within the dense hairs of the galls.

CULTURAL CONTROL

New plants may become infested in the nursery. Ensure that material used to establish new vineyards is free of mites and other pests or diseases. Removing galled leaves can control light infestations to some extent. Research conducted in the Okanagan Valley demonstrated that removal of infested leaves in May from Sauvignon Blanc vines did not alter yields or fruit quality.

CHEMICAL CONTROL

Erineum mite is seldom a problem in vineyards where sulphur is applied routinely for powdery mildew control.Applications of sulphur (e.g. Kumulus™ DF, or wettable sulphur) are most effective early in the season when galls are first being formed on new leaves. Sulphur sprays should not be used on Foch or *Vitis labrusca*varieties, as they are susceptible to leaf and fruit damage. Miticides (acaricides) such as Pyramite™ or Kelthane™ applied against spider mites and sprays of organophosphates (e.g. malathion and Sevin™) for control of leafhoppers will also reduce numbers of erineum mite somewhat.

GRAPE LEAF RUST MITE

(*Calepitrimerus vitis* (Nalepa)) Grape leaf rust mite, which belongs to the same family (Eriophyidae) as grape erineum mite, was identified for the first time on grapes in the Okanagan Valley in 2009. Growers are referred to the following source for additional information:Grapevine pests and their management, by A. Loch. Primefact 511, New South Wales Dept. of Primary Industries, NSW Australia.

IDENTIFICATION

A microscope is required to properly identify these 0.2 mm long pale white wormlike mites that closely resemble erineum or blister mites, but do not form leaf galls. They are much smaller than spider mites and elongate rather than round, tapering at the rear, and having two pair of legs near the head. Their presence is best determined by the bronzing of leaves during July and August that results from their feeding. Unlike spider mites that cause similar damage to leaves, grape leaf rust mites do not produce webbing.

LIFE CYCLE AND DAMAGE

Adult rust mites that spend the winter under bark or bud scales move in spring to the developing buds to feed and lay eggs. Large numbers of adults concentrated on the available green tissue in early spring can damage buds, resulting in deformed leaves, scarring of shoots, reduced shoot growth, and even bud death. Early season rust mite damage can be confused with cold injury or damage from other causes. Damage becomes less visible as the season progresses and shoots lengthen. Immature mites that hatch over a period of several weeks feed under the bud scales, eventually moving to feed on developing leaves. Unlike spider mites, rust mites infest equally the upper and lower leaf surfaces. Many generations are produced throughout the summer, and before leaves drop in the fall adult rust mites migrate to the cordon and trunk for the winter. Bronzing of leaves is most obvious during July and August, but damage to leaves is not thought to cause serious damage to the vines. Extensive bronzing of leaves is an indication that rust mites need to be controlled the following spring.

Fig. *Grape leaf rust mite damage to shoots.*

MONITORING AND SPRAY THRESHOLDS

There are no established thresholds for this pest. Control is indicated when a significant proportion of leaves become bronzed in late summer.

BIOLOGICAL CONTROL

As for erineum mite, western predatory mite, *Galendromus occidentalis*, other mite predators, and predacious thrips are often able to keep rust mites in check. To maintain healthy populations of beneficial insects and predatory mites, it is important whenever possible to limit the use of insecticides and to apply selective materials that are less toxic to predatory mites.

CHEMICAL CONTROL

A successful rust mite control program developed in Australia relies largely on spring applications of sulphur (e.g. Kumulus™ DF, or wettable sulphur) in high volume sprays during bud swell to woolly bud stage. Sulphur sprays should not be used on Foch or *Vitis labrusca* varieties, as they are susceptible to leaf and fruit damage. There are no miticides currently registered for the control of grape leaf rust mite.

GRAPE MEALYBUG

(*Pseudococcus maritimus* (Ehrhorn))

Mealybug were seldom a concern on grapes prior to the widespread use of synthetic insecticides. Individual bunches of table grapes that touched infested vines would occasionally become infested. Largely due to the detrimental effects of broad-spectrum insecticides on beneficial insects, damaging populations of mealybug have become more common in recent years. As for many other secondary pests, the most important management strategy is to apply chemicals against major pests (leafhoppers, cutworm) only when required and only to portions of the vineyard where control is warranted. Whenever possible, select pesticides that are least damaging to predators and parasites.

IDENTIFICATION

The grape mealybug (Homoptera: Pseudococcidae) has a flattened, oval pink body covered in a mealy white wax coating. It is somewhat segmented in appearance, but the divisions between head, thorax and abdomen are not distinct. Mature wingless females are about 5 mm long.

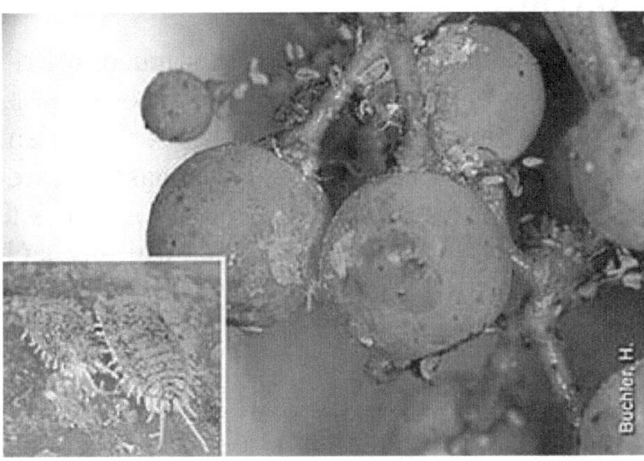

4

Grapes infested with grape mealybug. Grape mealybug

They have long waxy filaments along the edge of the body that are longest at the rear and become progressively shorter toward the front end. Large numbers of eggs are laid in cottony masses. The smaller, winged males have a pair of long, white anal filaments. Yellow to brown crawlers that emerge from the oval, orange eggs are not covered in wax. All stages are mobile.

LIFE CYCLE AND DAMAGE

Females move in late fall to old wood and lay overwintering eggs in cottony egg masses under loose bark. Crawlers, some of which might have hatched in fall and remained dormant throughout the winter, move to new shoots in spring. Grape mealybugs mature around mid-summer and produce a 2nd generation that matures in late autumn. They can occur on all plant parts but are more common in summer on leaves and new growth, usually hidden within the canopy.

Damage is similar to that for soft scale. Plant sap is removed during feeding and large amounts of honeydew foul the fruit and promote the growth of sooty mold fungus. Most fruit damage occurs during the development of the second brood in late summer. Mealybugs can infest grape bunches and are known to transmit certain strains of leafroll virus.

MONITORING AND SPRAY THRESHOLDS

To detect when crawlers are active, remove the loose bark in an area near the infestation and encircle the vine with clear tape applied sticky side out, or use double sided adhesive. Replace the tape every three or four days and check for the yellowish brown crawlers. A good magnifying glass or dissecting microscope is useful for this purpose. Unlike scale insects, grape mealybug prefer vigorous vines with thick canopies. The need to inspect vineyards for mealybugs can be based partly on past infestations, vigorous growth, and prior use of broad-spectrum insecticides.

BIOLOGICAL CONTROL

Parasitism of grape mealybug has not been studied in British Columbia. Research in California has shown that at least six species of parasitic wasps attack grape mealybug - late summer and fall rates of parasitism often exceed 90%. Predaceous midge larvae attack mealybug egg masses. Lacewings, several species of lady beetles, and many other generalist predators feed on adults. Mealybug and other homopteran insects (scale, whitefly, aphids, etc.) are susceptible to several diseases, including *Entomopthora* fungus.

CULTURAL CONTROL

Mealybugs prefer vigorous vines. Ensure adequate nutrition and moisture to produce an optimum crop, but avoid excessive vigour. Research in California has shown that fruit on cane-pruned varieties is less likely to suffer damage compared with spur-pruned vines. Clusters on cane-pruned vines hang farther away from the mealybug overwintering sites on old wood. Earlier varieties often escape damage as they are not exposed to honeydew to the same extent as later varieties. Heavy infestations of mealybugs and other honeydew-producing pests of grape are usually tended by ants that feed on the sugary excretion and protect the mealybugs from predation. Control of ants can often contribute to a significant decline in mealybug numbers.

For small backyard plantings with an isolated infestation, painting a sticky material around vine trunks and posts and ensuring that ants cannot gain access to the mealybugs often eliminates the problem. Trials in California have also shown that sprays directed against the Argentine ant, *Iridomyrmex humilis*, controlled damaging populations of the obscure mealybug, *Pseudococcus affinis*. Insecticides are not currently registered for the control of ants in B.C. vineyards.

CHEMICAL CONTROL

Malathion, Diazinon and Safer's Insecticidal Soap can be applied against the crawler stage, while Movento™ (spirotetramat) is active against all stages due to its good systemic activity.

SCALE INSECTS

About a dozen species of scale insects are known to attack grapes in various regions of the world, but only two are commonly found on grapes in British Columbia The European fruit lecanium scale (EFLS) and cottony maple scale (CMS) are both soft scales (Homoptera: Coccidae) that rarely cause economic injury to grapes. They are a greater concern in the production of table grapes. The large quantities of honeydew they produce supports the growth of sooty mold fungus that can leave fruit unmarketable. Two species of hard scale (Homoptera: Diaspididae), the oystershell scale,*Lepidosaphes ulmi* L., and San Jose scale, *Quadraspidiotus perniciosus* Comstock, can occasionally be found attacking 2 year old wood of weakened vines. Hard scales do not produce large amounts of honeydew and are currently of little importance to the B.C. grape industry.

IDENTIFICATION

Female EFLS, also known as brown apricot scale, when fully grown have a chestnut brown, smooth and slightly shiny protective shell that is around 5 to 8 mm long and slightly longer than wide. Certain races in some regions reportedly produce male scales, but most females reproduce without mating. The small nymphs or crawlers that hatch from the small oval, pearly white eggs are yellow to orange in colour. Mature female CMS are similar in size, shape and colour to EFLS, except that a mass of cottony material containing the eggs, up to 1,000 per sac, extrudes from the rear end. The much smaller winged males produced in late summer have no mouthparts and live for only 1 or 2 days. Males mate with the immature females. Mature scales of both species become brittle after the eggs are laid in mid-summer and can easily be detached from the one to three year old wood with a fingernail. The smaller hard scales (1-2 mm), which are much longer than wide, remain firmly attached to the canes even after death.

Adult female cottony maple scale with cottony egg masses being tended by ants. Immature cottony maple scale

European fruit lecanium scale

LIFE CYCLE AND DAMAGE

EFLS and CMS overwinter as partly grown scales on new canes. Growth resumes in spring and females reach maximum size around early summer when they begin to lay eggs. The newly hatched scale called crawlers hatch and disperse during July and August to the undersides of leaves where they settle and feed until early fall, when they migrate to the current year's wood and develop a hardened protective shell for the winter.

Both species of soft scale feed on the phloem of grapes and a large number of other woody plants, including many fruit trees and ornamentals. Scales are not usually abundant on grapes grown in British Columbia, and direct feeding damage from the removal of plant fluids is rarely a cause for concern. They appear to prefer weak vines, however, and the additional stress might be sufficient to hasten death. Of greater concern, they produce vast amounts of honeydew that can turn grape bunches into sticky, unmarketable messes. Additionally, honeydew supports the growth of sooty mold fungus which further disfigures the fruit.

MONITORING AND SPRAY THRESHOLDS

Because scale infestations often occur on only a single vine or a few vines in a small area, they can be difficult to detect. Partly-grown scales might be observed during pruning, or watch for honeydew on leaves and fruit during spring and early summer. Ants feed on the honeydew and will protect the scale from predators. They can often be seen in large numbers hurriedly running up and down posts or vine trunks to and from infested canes.

To detect when crawlers are active, remove the loose bark in an area near the infestation and encircle the vine with clear tape applied sticky side out or use double sided adhesive. Replace the tape weekly and check for the orange coloured crawlers. A good magnifying glass or dissecting microscope is useful for this purpose

There are no established thresholds for scale infestations. Small numbers of scale on a healthy vine are unlikely to cause damage; treatment is usually only required when the excretion from large infestations fouls table grapes.

BIOLOGICAL CONTROL

Biological control of scale on grapes in British Columbia has not been investigated in any detail, but on other crops and in other regions of the world they are known to be heavily parasitized by several species of wasps and flies. They are preyed upon by lacewings, lady beetles, and other generalist predators. English sparrows and several species of warblers that feed on fully grown soft scale have been known to curtail infestations.

CULTURAL CONTROL

Healthy vines of moderate vigour are less susceptible to scale infestations and are better able to withstand damage. Routine pruning removes many scales and a small infestation can be pruned out in spring or early summer.

CHEMICAL CONTROL

Movento™ (spirotetramat) is registered for the suppression of lecanium scale. Malathion and insecticidal soap will also help control the active crawler stages in summer. Repeat applications might be required as long as crawlers are active.

SNAILCASE BAGWORM

(*Apterona helix* Siebold)

IDENTIFICATION

This introduced moth (Family Psychidae) remains throughout most of its life within a snail-like case made from particles of soil and sand bound together with silken threads. When fully grown the coiled case is approximately 4 mm in diameter. There are no males and the wingless female moths remain inside the case to deposit their eggs without having mated (parthenogenetic reproduction). Large aggregations often form on posts, tree trunks, or vine trunks in late summer.

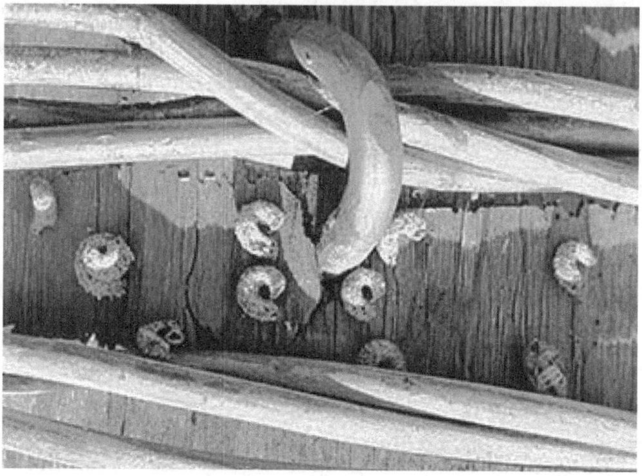

Snailcase bagworm larvae on wires Snailcase bagworm larvae on post

LIFE CYCLE AND DAMAGE

Young caterpillars overwinter within the case formed by the mother. They drop to the ground in spring, form C-shaped cases, and begin to feed on a wide range of weeds, native plants and cultivated crops. Damage consists of small excavated holes on the leaf surface. As the larva grows it eventually expands its case into the typical snail-like, helical shape. In mid-summer the fully grown larvae will climb vines, posts or buildings, attach themselves firmly with silk, and pupate within the case. Several weeks later the females emerge, lay eggs, and then exit the case to die. The eggs hatch in late summer and the small first instar larvae spend the winter within the protective cases. There is only one generation per year.

MONITORING AND SPRAY THRESHOLDS

Larvae of the snailcase bagworm cause minimal feeding damage to leaves and they are not considered economic pests of wine grapes. The large aggregations are easily detected on buildings, posts, wires, and the trunks of vines. If contamination of table grapes is a concern, inspect vines and apply control measures before the overwintering aggregations form in late summer. The non-feeding overwintering stage, secure within the case, is not susceptible to insecticides.

BIOLOGICAL CONTROL

Because it has not yet been adequately studied, natural enemies of this pest of grapes in British Columbia are unknown. The protective case would provide some protection from certain generalist predators, but predators, parasites and diseases undoubtedly take their toll.

CULTURAL CONTROL

Snailcase bagworm prefer moist areas with thick vegetation. Avoid excessive irrigation and drain areas that puddle or remain damp. Proper drainage will also help reduce grapevine root diseases. Occasional discing or mowing of areas of the vineyard where bagworm are a problem can help reduce infestations.

CHEMICAL CONTROL

Dipel™ (*Bacillus thuringiensis*) is registered for snailcase bagworm control on grapes. Sprays applied for the control of other insects will provide some protection. For this reason, they are more common in organic vineyards or conventional vineyards with low chemical inputs.

EUROPEAN EARWIG

(*Forficula auricularia* L.)

The European earwig is both a beneficial insect and a pest of table grapes. All stages feed on small, soft-bodied insects such as small leafhopper nymphs and will scrape leafhopper eggs off of leaves. In fall they feed to some extent on damaged grapes and their presence in table grapes is not appreciated by some consumers. On grapes produced for wine, they should be considered important generalist predators.

IDENTIFICATION

Adult earwigs (Order Dermaptera) are dark brown, about 1.5 cm long, with a pair of forcep-like appendages (cerci) at the rear of their elongate bodies. The anal forceps of females are long and straight, while those of males are curved. Earwigs seldom fly and the membranous hind wings usually remain folded up under the short leathery forewings, exposing the abdomen.

The four nymphal stages resemble the adults except that they are smaller and wingless. Earwigs are largely nocturnal and hide during the day, often in large aggregations, in cracks and crevices, or under bark. When disturbed they move quickly.

LIFE CYCLE AND DAMAGE

Adults overwinter in the ground in earthen cells. Clutches of eggs laid in cells in the top 5 cm of soil in the fall and spring are opened by females in spring to release the young nymphs. Adults are semi-social and tend their young for a period of time after birth; nymphs later disperse to secure locations under rocks, in crevices, or under bark. Adults develop by midsummer and there is only a single generation per year. Earwigs are nocturnal and feed at night on damaged fruit, leaves, small soft bodied insects, and other foodstuffs. They rarely damage grape leaves.

MONITORING AND SPRAY THRESHOLDS

Rolled up newspaper or cardboard, or flat boards placed on the ground, can be used to monitor earwigs. It might be possible to use these same refuges to enhance their numbers in wine grapes.

BIOLOGICAL CONTROL

Birds, large spiders, wasps and ground beetles prey on earwigs. Earwigs, in turn, feed on the eggs and small nymphs of leafhoppers and on other soft bodied insects.

CULTURAL CONTROL

Tillage in spring and fall will destroy some of the earthen cells. For table grapes, mowing, tillage, removal of loose bark from vine trunks, and sanitation will remove hiding places and help reduce earwig numbers. They can be kept out of small plantings of table grapes by applying sticky tape or commercial Stickum around trunks and posts. Washing table grapes after harvest will help dislodge many earwigs. Earwigs are less likely to seek refuge in table grape varieties with loose, open fruit bunches.

CHEMICAL CONTROL

Earwigs are sensitive to most insecticides and sprays applied for the control of other grape pests will usually also provide control of earwigs. Commercial earwig baits are also available.

CLICK BEETLE

(Coleoptera: Elateridae)

Click beetles were formerly considered to be important pests of grapes in British Columbia, but extensive study in Okanagan vineyards has shown that damage attributed to click beetle adults was in fact caused by half grown cutwormlarvae. Several species will feed occasionally on buds of grapes, but control is usually not warranted. Click beetles are also not considered pests of grape in Washington State.

IDENTIFICATION

Click beetle larvae, called wireworms, have long, slender, hard bodies that are a light tan to a light orangey-brown in colour, and appear somewhat shiny and polished. Their short legs and cylindrical slender bodies allow them to burrow through the soil to feed on the roots of plants. Adult beetles have elongate, somewhat flattened bodies usually in shades of brown or black. The segment behind the head has small projections that point backward on each side of the body. Adults of most species are around 1 to 1.5 cm in length, but some are significantly larger. When adults are placed upside down in the palm of a hand or gently held between finger and thumb they will rapidly flex their bodies at both ends and attempt to right themselves or escape with a characteristic clicking action.

LIFE CYCLE AND DAMAGE

Adults that have spent the winter under debris and leaves emerge with the arrival of warm weather in spring to feed on the tender buds and leaves of a wide range of plants, including occasionally grapes. They are often quite common on flowers where they feed on soft tissues and pollen. After feeding for a short time, females begin to deposit eggs into the soil. In two or three weeks the eggs hatch into small slender 'worms', the larvae, that develop in the soil throughout the summer. Click beetles can persist for a long time without feeding and they often require more than one year to complete development to the adult stage. Larvae are not known to damage the roots of grapes to any extent, preferring to feed on the roots of weeds or grasses. As for cutworm larvae, feeding of adults on buds of grape causes the buds to die. Yields can be reduced and ripening of fruit can be uneven and delayed. Little damage occurs after bud break.

MONITORING AND SPRAY THRESHOLDS

Adult click beetles feed on buds so infrequently that monitoring for these pests is generally not warranted. The presence of adult click beetles in a vineyard does not indicate feeding or correlate with damage. There are no established thresholds in British Columbia relating click beetle numbers to bud damage. To determine if bud damage is due to click beetles or cutworm larvae, it is best to inspect vines with the aid of a flashlight soon after dark on warm spring nights.

BIOLOGICAL CONTROL

Little information is available on the biological control of click beetles in British Columbia.

CULTURAL CONTROL

As for climbing cutworm, the presence of broadleaf weeds in spring helps reduce damage to grapes.

CHEMICAL CONTROL

Chemical control of click beetles is generally not warranted and none are registered specifically for adults on bearing vines. Sprays for cutworm larvae should help control adult click beetles.

7

Grape Production

GRAPE PRODUCTION IN INDIA

Grape cultivation is one of the most remunerative farming enterprises in India. Famous Indian medicine scholars, Sasruta and Charaka in their medical treatises entitled *'Sasruta Samhita'* and *'Charaka Samhita'*, respectively, written during 1356-1220 BC, mentioned the medicinal properties of grapes. Kautilya in his *'Arthashastra'* written in the fourth century BC mentioned the type of land suitable for grape cultivation. Native spp. resembling *Vitis lanata* and *Vitis palmata* grow wild in the northwestern Himalayan foothills. Indigenous varieties known as 'Rangspay', 'Shonltu White' and 'Shonltu Red' are grown in Himachal Pradesh even today.

Cultivated grapes are believed to have been introduced into the north of India by the Persian invaders in 1300 AD, from where they were introduced into the south (Daulatabad in Aurangabad district of Maharashtra) during the historic event of changing the capital from Delhi to Daulatabad by King Mohammed-bin-Tughlak. Ibn Batuta, a Moorish traveller who visited Daulatabad in 1430 AD, reported to have seen flourishing vineyards in south India. Grape was also introduced in the south into Salem and Madurai districts of Tamil Nadu by the Christian missionaries around 1832 AD, and into Hyderabad province by HEH, the Nizam of Hyderabad in the early part of the 20th century. From Delhi, Daulatabad, Madurai, Salem and Hyderabad, grape cultivation spread to different parts of the country.

PRESENT STATUS OF GRAPE CULTIVATION IN THE COUNTRY

Grape is grown under a variety of soil and climatic conditions in three distinct agro-climatic zones, namely, sub-tropical, hot tropical and mild tropical climatic regions in India.

Sub-tropical Region: This region covers the northwestern plains corresponding to 28° and 32° N latitude including Delhi; Meerut district of Uttar Pradesh; Hissar and Jind districts of Haryana; and Bhatinda, Ferozpur, Gurdaspur and Ludhiana districts of Punjab. Vines undergo dormancy and bud break starts in the first week of March while the rains arrive in the first week of June, and therefore, only 90-95 days are available from the initiation of growth to harvest. Consequently, 'Perlette' is the only early ripening variety grown in this region. Rain damage is a problem with Thompson Seedless in this region. Single pruning and a single harvest is the accepted practice here.

Hot Tropical Region: This region covers Nashik, Sangli, Solapur, Pune, Satara, Latur and Osmanabad districts of Maharashtra; Hyderabad, Ranga Reddy, Mahbubnagar, Anantapur and Medak districts of Andhra Pradesh; and Bijapur, Bagalkot, Belgaum, Gulberga districts of northern Karnataka lying between 15° and 20° N latitude.

This is the major viticulture region accounting for 70 percent of the area under grapes in the country. Vines do not undergo dormancy and double pruning and a single harvest is the general practice in this region. Maximum and minimum temperature is 42°C and 8°C, respectively. The major problems in this region are soil and water salinity and drought. Berry growth is impaired and in certain locations pink blush sometimes develops on green berries due to temperatures that drop to a low of 8°C. Thompson Seedless and its clones (Tas-A-Ganesh, Sonaka), Anab-e-Shahi, Sharad Seedless and Flame Seedless are the varieties grown in this region.

Mild Tropical Region: An area covered by 10° and 15° N latitude including Bangalore and Kolar districts of Karnataka; Chittoor district of Andhra Pradesh and Coimbatore; and Madurai and Theni districts of Tamil Nadu fall in this region. Maximum temperatures in a year seldom exceed 36°C, while the minimum is about 12°C. Principal varieties are Bangalore Blue (Syn. Isabella), Anab-e-Shahi, Gulabi (Syn. Muscat Hamburg), and Bhokri. Thompson Seedless is grown only with limited success. Except for Thompson Seedless, two crops are harvested in a year.

Vinifera varieties susceptible to mildew suffer losses due to unprecedented rains during flowering and fruit set in both hot and mild tropical regions.

Area and production of different varieties of grapes in India is as follows:

Variety	Area (ha)	Production (t)
Anab-e-Shahi (white, seeded)	3,000	135,000
Bangalore Blue Syn. Isabella (black, seeded)	4,500	180,000
Bhokri (white, seeded)	500	15,000
Flame Seedless (red, seedless)	500	10,000
Gulabi Syn. Muscat Hamburg (purple, seeded)	1,000	30,000
Perlette (white, seedless)	1,500	60,000
Sharad Seedless - A mutant of Kishmish Chorni (black, seedless)	1,000	20,000
Thomson Seedless and its mutants (white, seedless)	22,000	550,000
Total	**34,000**	**1,000,000**

Approximately 85 percent of the total production, irrespective of the variety, is consumed fresh. About 120,000 tonnes of Thompson Seedless and its mutants, namely, Tas-A-Ganesh, Sonaka and Manik Chaman are dried for raisins. Some 20,000 tonnes of Bangalore Blue are crushed to make juice, and

10,000 tonnes of Bangalore Blue, Cabernet Sauvignon, Chenin Blanc, Chardonnay, Merlot, Pinot Noir and Uni Blanc are crushed to process into wine.

PRODUCTION OF PLANTING MATERIAL

Vines are raised on their own roots in India. Due to the non-prevalence of Phylloxera or nematodes, rootstocks are not employed, but in recent years, the 'Dogridge' rootstock is being employed to combat soil and water salinity problems.

Multiplication on Own Roots

Grapes are multiplied exclusively by the rooting of hardwood cuttings. No Government agency is involved in the multiplication and supply of rooted cuttings. Growers themselves obtain the hardwood cuttings from elite vineyards and raise their own nurseries. Well matured canes obtained in September/ October are selected. Cuttings of 4 nodes each with a thickness of 8 to 10 mm are made from the selected canes.

The fresh cuttings are soaked in running water for 24 hours to leach out the water-soluble rooting inhibitors. The basal parts of cuttings are then dipped in a 2,000 ppm strong IBA solution for five minutes before planting. It is also a practice to plant the cuttings *in situ* when three to four cuttings prepared and treated as above are planted at each spot in the main field. Soil drenching with chlorophyriphos 0.1 percent is a practice to safeguard the cuttings against termite damage.

Raising on Rootstocks

Hardwood cuttings of the 'Dogridge' rootstock are subjected to rooting, preferably in polybags of 15 x 25 cm. Rooted cuttings of this rootstock are planted in the main field during February-March. The desired scion variety is then grafted/budded on the rootstocks in the field by wedge grafting/chip budding. Wedge grafting is more common and the best time for the operation is September-October, while June-July is the suitable time for chip budding.

ESTABLISHMENT OF VINEYARDS

Land Preparation and Vine Establishment

The land is tilled and laid into plots of 120 m x 180 m separated by 3 m wide roads. Land within a plot is levelled perfectly to have a gradient of less than 1 percent in any direction to ensure uniform discharge of water through the emitters of drip irrigation systems.

Trenches of 75 cm width, 75 cm depth and 118 m length in a north-south direction with a gap of 3 m between trenches are opened with heavy machinery. They are closed with topsoil, up to a height of 45 cm after 15 days exposure to sun. The remaining gap is filled with a mixture of soil, cattle manure, single superphosphate, sulphate of potash and micro-nutrients. Usually, 50 kg of cattle manure, 2.5 kg of superphosphate, 0.5 kg of sulphate of potash and 50 g each of $ZnSO_4$ and $FeSO_4$ are added to the soil for every running meter length of the trench.

Planting Season

The best season for planting the rooted cuttings of cultivated varieties in the main field is September-October whereas for rootstocks it is February-March.

CARE AND MANAGEMENT OF VINEYARDS

Training of Vines

Many training systems are in vogue in India, but the most popular are Bower, Telephone and Flat Roof Gable systems.

Bower System: Owing to the high productive potential, bower was a very popular system of training in the past. It is highly suited for vigorous varieties like Anab-e-Shahi, Bangalore Blue and Gulabi. But in varieties like Thompson Seedless and Tas-A-Ganesh where vine vigour and excessive foliage density affects the productivity adversely, this system is not popular.

Telephone System: T-trellis is used in this system of training. With three top wires and 'T' shaped supports, the trellis looks like a telephone pole and wires and hence the name.

This system is followed for moderately vigorous varieties like Thompson Seedless and other seedless cultivars in about 25-30 percent of the vineyard area in Maharashtra. Yields in this system are less than the bower. In very hot and dry places, sunburn of the berries and of the arms are experienced in summer.

Flat Roof Gable System: Combining the advantage of bower and the extended Y systems and eliminating their disadvantages, an inter-connected Y trellis forming a flat roof gable is being adopted. This system is particularly followed for vigorous vines (vines grafted on rootstocks). The bunches are protected from direct sunlight and well exposed to sprays of pesticides. The clusters hang within the reach of the worker of an average height. Owing to these advantages, this system is gaining popularity among the growers in Maharashtra, Andhra Pradesh and Karnataka.

Pruning of Vines

Three distinct pruning practices are in vogue in relation to cropping in the three grape growing regions of the country. In the sub-tropical region, vines are pruned only once in December and the crop is harvested once. Half of the canes are pruned to renewal spurs and the rest to fruiting canes (3-4 nodes for Perlette).

In hot tropical regions, vines are pruned twice but only one crop is harvested. All canes in a vine are pruned back to single node spurs in March-May to develop canes and the canes are forward pruned in October-November for fruiting. The number of nodes retained on a cane varies with the variety and cane thickness. There is no scope to prune earlier than October and later than November due to unfavourable weather conditions.

In the mild tropical region, vines are pruned twice and the crop is harvested twice. In varieties like Gulabi and Bangalore Blue, which are fairly resistant to rain damage and in which fruit bud differentiation is not impaired by cloudy weather and rains, pruning is done at any time of the year. As a result, five crops are harvested every two years.

Application of Manure and Fertilizers

As vineyard soils are either sandy loams or heavy clays, the usage of organic manure has assumed high importance in India. A standard dose of 500:500:1000 kg of N, P_2O_5 and K_2O per hectare is followed in light sandy soils, while 660:880:660 kg are applied for heavy clay soils. The annual dose is fixed based on the petiole analysis carried out at 45 days after spur pruning. While 40 percent of the annual dose is given through organic sources, 60 percent is given as inorganic fertilizer. Calcium ammonium nitrate is usually not used. Sulphate of potash is the only source of potash used in place of muriate, particularly in heavy clay soils. Recently application of soluble fertilizers through drip irrigation is picking up. 40 percent of N, 50 percent of P_2O_5 and 33 percent of K_2O of the annual dose is given during the growth season and the rest in the fruiting season.

Weeding

Weeds between the rows of vines are removed mechanically by tractor drawn implements. Within the rows, weeds are manually hoed and removed. Sometimes the post-emergent weedicides, mainly glyphosate at about 2.0 kg/ha or paraquat at about 7.5 kg/ha is sprayed in fully grown vineyards.

Supplementary Irrigation

Since grapes are grown in areas where the evapotranspiration exceeds the precipitation, irrigation is essential. Less than 10 percent of the vineyard areas are surface irrigated, while the rest is irrigated by drip systems. Water requirement is calculated based on the pan evaporation using 0.8 as the crop factor. Water is applied at different rates at different stages of vine growth and berry development.

Pests and their Management

The important pests of grapes in India are, flea beetles, thrips, mealy bugs and leaf hoppers.

Flea beetles: The adult beetles scrape the sprouting buds and eat them up completely after each pruning. Damaged buds fail to sprout. Insecticides like carbaryl at 0.15 percent, quinolphos at 0.05 percent, dichlorvas at 0.1 percent or phosalone at 0.05 percent are sprayed from the fourth day until the emergence of leaves.

Thrips: Thrips attack the ovaries of flowers and newly set berries and suck sap from them. The affected berries develop a corky layer and become brown on maturity.

Scab formation on the berry surface is also due to thrip damage to the ovaries/young berries. Such berries are not suitable for marketing. Thrips are effectively controlled by spraying phosphamidon at 0.05 percent, carbaryl at 0.125 percent, phosalone at 0.05 percent or malathion at 0.05 percent. Prophylactic sprays of insecticides against thrips are given once in five days from the initiation of bloom to berry set.

Mealy Bugs: Mealy bugs are the most serious and problematic pests of grapes in India. Nymphs and adults suck sap from the tender shoots resulting in crinkling and stunting of the new shoots. They excrete honey on leaves and berries and sooty mold develops on the honey. Mealy bug infected bunches are unfit for marketing. Yield losses can be up to 50 percent due to mealy bug damage. Mealy bugs are hard-to-kill insects and the package of practices for their control in India is as follows:

i) Avoid spraying broad-spectrum insecticides particularly synthetic pyrethroids.
ii) Spray only dichlorvas at 0.1 percent mixed with neem oil 0.2 percent or tridemorph at 0.1 percent.
iii) Release *cryptolaemus montrozieri* beetles at 8,000-10,000 per hectare when the berries start softening. It is better to release a mixed population of grubs and adults rather than only adults.

Leaf hoppers: This pest has assumed serious proportions in all grape growing regions of India in recent years. The adults and young nymphs of hoppers suck sap exclusively from the lower side of the leaves. Carbaryl at 0.15 percent, fenitrothion at 0.04 percent, phosalone at 0.05 percent or quinalphos at 0.05 percent are sprayed to control this pest. A mixture of quinalphos at 0.05 percent and phosalone at 0.05 percent is more effective on the nymphs while tridemorph at 0.1 percent only is effective on the adults.

Diseases and their Management

The important grape diseases are anthracnose, downy mildew, powdery mildew and bacterial leaf spot. In recent years, *Alternaria* is also becoming a serious pathogen.

Anthracnose is prevalent in all grape growing regions of the country. The disease is characterized by small light brown or greyish black lesions on tender shoots, young leaves, flowers and young berries. Bordeaux mixture at 0.8 percent, copper oxychloride at 0.25 percent or carbendazim at 0.1 percent are used to control this disease.

Downy mildew is the most devastating disease of grapes in the tropical region of the country. The disease mainly appears on the leaves, but also attacks the flower clusters and young fruits.

The losses are very high when it attacks the clusters before fruit set. Entire clusters decay, dry and drop down. Properly neutralized Bordeaux mixture at 1 percent, copper oxychloride at 0.2 percent, Mancozeb at 0.2 percent, metalaxyl (Ridomil Mz at 0.2 percent) or Phosethyl-Al (aliettle at 0.2 percent) are used against this disease.

Powdery mildew is prevalent in all the grape growing regions. It is next in importance to downy mildew in its devastating severity. The disease is characterized by the presence of white powdery (ash like) coating in patches on both sides of the leaves, young shoots and immature berries. Powdery mildew is controlled easily by wettable Sulphur formulations.

A wide range of fungicides, namely, Calaxin at 0.07 percent, Karathane EC at 0.04 percent, Myclobutanil (Systhane at 0.05 percent), Triademifon (Bayleton at 0.1 percent) and Penconazol (Topas at 0.025 percent) are used to control this disease.

Bacteria infects leaves, shoots and berries. The symptoms appear as minute water soaked spots on the lower surface of the leaves, especially along the main and lateral veins. Mostly these spots coalesce and form larger patches. Severely infected leaves give a blighted appearance. Streptocyclin at 500 ppm is used as a prophylactic spray, while Bordeaux mixture at 0.8 percent or copper oxychloride at 0.15 percent is used to check its spread.

Physiological Disorders

Physiological disorders associated with high temperature and low atmospheric humidity in the hot tropical region are dead arm and trunk splitting. Salinity injury is common in Maharashtra and north Karnataka. Other physiological disorders are cane immaturity, water berries, cluster tip wilting, shot berries, uneven ripening and post-harvest berry drop. The eco-physiological disorders are *'coulure'*, blossom-end rot, pink berry syndrome, berry cracking and rotting.

Quality Improvement

Shoot and Cluster Thinning: Only one or two clusters are retained per cane depending upon the density of the latter. Irrespective of the number of clusters, only the apical two or three shoots are retained. In vines trained to the flat roof gable, individual shoot length is encouraged rather than the total canopy size for preventing sunburn of the berries.

Production of Loose Clusters: Pre-bloom GA sprays of 10 ppm and 15 ppm are given respectively on the 11th to 14th day after bud break for cluster elongation. Rachides of the clusters are trimmed to retain 8-10, depending on the number of leaves available per cluster. Clusters are dipped in GA solution of 30-40 ppm when 10-20 percent of the flowers open in each cluster for berry thinning.

Increasing Berry Size: Manual means are used to supplement chemical thinning to ensure adequate berry thinning and improve the quality of grapes. Approximately 90-120 berries are retained per cluster depending upon the number of leaves available to nourish it at 8-10 berries per every leaf depending on its size. Clusters are dipped in GA solution of 40-50 ppm concentration once at 3-4 mm size of the berries and again at 7-8 mm size. When berry diameter is to be increased to more than 16 mm, clusters are dipped in a mixture of 10 ppm BA + 25 ppm GA or 2 ppm CPPU + 25 ppm GA or 1 ppm brassinosteroid + 25 ppm GA instead of GA alone at these two stages.

In addition to the treatment with growth regulators, berry size and crispiness are increased by girdling. The width and depth of girdling are 1-1.5 mm. Girdling is done at 4-5 mm diameter of the berries.

Increasing the TSS Content: Berry thinning and cluster thinning to maintain adequate leaf/fruit ratio (5 cm^2), while girdling will ensure a TSS content of 20°B.

HARVESTING AND YIELDS

Approximately one million tonnes of grapes are harvested annually in India. Grape is harvested almost all the year round. If not all the varieties, one or more varieties are always available at any given time of the year. Period of harvest and yield of different varieties is given below.

Variety	Yield (t/ha)		Period of Harvest
	Average	Potential	
Anab-e-Shahi	45	90	February-May, July, November-December
Bangalore Blue	40	60	January-March, June-December
Bhokri	30	50	November-December, June-July
Gulabi	30	50	January-March, June-December
Perlette	40	50	June
Thompson Seedless and other seedless varieties	25	50	January-April

However, the major proportion of produce, mainly of Anab-e-Shahi, Thompson Seedless and its clones, is harvested during March-April from the hot tropical region, which contributes more than 70 percent of the total harvest.

The productivity of grapes in India is very high, particularly in the Hyderabad region. Yields as high as 100 t/ha in Anab-e-Shahi and 75 t/ha in Thompson Seedless were recorded in this region. However, quality of grapes is usually poor as a result of high yields.

MARKETING

More than 80 percent of the total production is consumed as table grapes in India, and more than 70 percent of the total production is harvested in March-April, but the cold storage facilities are inadequate. Therefore, market gluts and fall of prices of grapes in March-April are common. Approximately, 2.5 percent (22,000 t) of fresh grapes are exported to the Middle East and European countries. The rest of the produce is marketed within the country. Grapes are exported through three different agencies viz., Grower Exporters, Growers' Cooperatives and the Trader exporters. These agencies have established their own facilities for pre-cooling and cold storage in the vicinity of major production sites.

POTENTIAL FOR GRAPE PRODUCTION DEVELOPMENT

India has the distinction of achieving the highest productivity in grapes in the world, with an average yield of 30 t/ha.

a) Sustaining productivity and minimizing risks in grape cultivation is possible because of the availability of a variety of agro-climatic regions suitable for grape cultivation for table, raisin and wine grapes.
b) Technologies to achieve high productivity are currently available.
c) Scope for double cropping in certain regions and harvesting round the year in certain varieties is practically feasible.
d) Technologies to produce export quality grapes and quality raisins are available.

CONSTRAINTS IN GRAPE PRODUCTION DEVELOPMENT

Although grape cultivation is considered as highly remunerative, the area under grapes is confined to only 34,000 hectares due to the following constraints.

a) Heavy initial investment for establishing a vineyard.
b) High recurring costs in vineyard management.
c) Narrow variety base and lack of diversity in utilization of the germplasm available in grape growing countries.
d) High risk of losing the crop due to unprecedented changes in weather.
e) Soil and water salinity in Maharashtra and drought in the hot tropical areas.
f) Short period available for ripening in the north.
g) Very low proportion of export quality grapes.
h) Wine is not a popular drink at present.
i) Marketing problems in table grapes.

GOVERNMENT POLICIES AND PLANS FOR RESEARCH AND DEVELOPMENT OF GRAPES

The Government of India is supporting the grape industry of the country in the following ways:

a) Encourage and support the farmers for establishing the vineyards and installing drip irrigation systems by providing soft loans and subsidies.
b) Provide research support to sustain the productivity of grapes under adverse situations.
c) Promote and support the export of fresh grapes by training the growers and providing soft loans and subsidies for pre-cooling and cold storage facilities.

Research on grapes is carried out by the Indian Council for Agricultural Research (ICAR) Institutes and State Agricultural Universities at different centres under the All India Coordinated Research Project on Grapes. The National Research Centre for Grapes (ICAR) located at Pune, Maharashtra is the focal point for conducting and coordinating the research activities on grapes throughout the country.

CONCLUSIONS

Grape is cultivated over an area of 34,000 hectares with an annual production of 1,000,000 tonnes. Although, the returns per unit area of land are very high with grape cultivation, the area under grapes is not expanding fast owing to the high initial cost of establishing the vineyards and high recurring cost of production. The risk of losing a crop due to unprecedented changes in weather is also very high. Since the highest productivity in grapes has been achieved, efforts are needed to extend grape cultivation to newer areas. Soil and water salinity and drought are the impediments in this direction, for which suitable rootstocks are to be identified.

There is a need to diversify the uses of grapes. Currently more than 80 percent of the produce is used for table purposes. The major bulk of the produce is harvested in March-April, but as cold storage facilities are currently inadequate there are frequent market gluts. Diversification of uses as wine/ juice and export of table grapes can ease the marketing problems. Maintenance of quality of table grapes by crop regulation is the priority consideration to increase exports. For the survival of the grape industry in India, the produce should be quality and cost competitive. Future efforts are to be concentrated in this direction.

GRAPE PRODUCTION IN JAPAN

Several species of the genus *Vitis*, such as *Vitis coignetiae* Pulliat., *V. flexuosa* Thumb., and *V. amurensis* Rupr., are native to Japan. However, they have not been brought into cultivation or for breeding. The first grape culture in Japan dates back to the 12th century, when a wild grapevine bearing good fruit bunches was found in Yamanashi. The vine was apparently not native to the land, suggesting that its ancestor might have been imported from China in the 8th century. The progenies from the vine adapted to the environment of that district, and the cultivar 'Koshu' was thus released. In the 17th century, 'Koshu' grapes became famous for their excellent fruit quality. In the late 19th century, the Government actively imported numerous species including *V.vinifera, V.labrusca,* and *V.labruscana* from the USA and European countries. However, most *vinifera* grapes could not be cultivated successfully because of the spread of diseases. Only a few *vinifera* cultivars, such as 'Muscat of Alexandria,' could grow normally with the aid of glasshouses. The vines of both *V.labrusca* and *V.labruscana*, on the other hand, grew successfully. 'Catawba,' 'Campbell Early,' and 'Delaware' were planted in Yamanashi, Osaka, and Okayama prefectures in the early 20th century.

Grape growers tried to breed new good quality table grapes suitable for the humid conditions of Japan. Some cultivars, such as 'Kyoho,' 'Muscat Baily A,' 'Neo Muscat,' and 'Hiro Hamburg,' were identified from the crosses between *V. vinifera* and *V. labruscana* or between *vinifera* grapes in the middle 20th century. These cultivars, however, could not be cultivated successfully because of poor berry set and susceptibility to fungal attack in those days.

The cultivation of grapes grew rapidly from 1960 to 1980. Success in the production of seedless 'Delaware' grapes by GA treatment further accelerated the spread of this cultivar. The cultivar 'Kyoho,' famous for its large sized berries and excellent taste, was also planted throughout the country except in Hokkaido, the northernmost island. Another contribution in the development of grape growing was the application of plastic films, which greatly decreased the spread of diseases and increased both yield level and fruit quality. The practice of forcing culture by heating the vines in plastic houses also spread. Recently, propagation of virus-free vines has been developed using the technology of *in-vitro* cultures of heat-treated stem apices. Vine growth and fruit quality have also been significantly improved. This has resulted in the spread of virus-free vines in most newly established vineyards that are suitable for replanting. The main producing areas have been established in the north, central, and western districts of the country.

Viticulture in Japan started from adoption of the variety 'Koshu' (*Vitis vinifera* var. *orientalis*), which was discovered in Yamanashi in 1186. Thereafter, European and American grape cultivars were introduced in the 1800's. Most *V. vinifera* grapes from Europe could not grow well because of serious damage by Phylloxera and many other diseases. On the other hand, American cultivars derived from *V. labrusca* were suited to the Japanese environmental conditions. 'Delaware,' 'Campbell Early,' 'Niagara,' and 'Concord' became popular at the beginning of the 20th century. Thereafter, the popularity of 'Kyoho' increased rapidly in the 1970's.

It is important to develop new grape cultivars that are well suited to the Japanese environmental conditions and consumers' preferences. Grape-breeding programmes in Japan that originated in the 1920's aimed at these objectives.

More than 200 grape cultivars have been developed so far. Thus, conventional techniques by cross-hybridization have contributed to the development of the grape industry. However, when using this approach, not only desirable but also unfavourable characteristics are inherited. Recent advances in biotechnology offer many strategies, such as gene transformation and molecular markers for breeding, and may help overcome the limitations of conventional techniques.

PRESENT STATUS OF GRAPE CULTIVATION IN JAPAN

Grape production has been undertaken in all four main islands in Japan from north to south. The major producing regions are Yamagata, Yamanashi, and Nagano prefectures in the northern and central parts, and Okayama and Fukuoka prefectures in the southern part of Japan. In 1997, the total grape production was 250,900 metric tonnes and the vineyard area was 22,800 hectares. Most vineyards in Japan produce table grapes. In 1997, table grapes made up 87.4 percent of the total production, while 11 percent and 1.4 percent went to the production of wine and juice, respectively.

Table. Grape Production and the Proportion of Table Grapes and Grapes for Processing (1997)

Product	Production (tonnes)	Proportion of total production (%)
Table Grape	219,300	87.4
Wine	27,700	11.0
Juice	3,500	1.4
Canned Fruit	400	0.1
Total Production	250,900	100

'Kyoho,' a tetraploid, which is one of the main cultivars in Japan, produces large sized berries exceeding 15 g with strong sweetness (18-20 Brix) and good flavour. Another major cultivar is 'Delaware', which is one of the *labrusca* cultivars introduced from the USA in the late 19th century. This is the first cultivar in which the production of seedless berries by GA (gibberellic acid) treatment was successful. Main cultivars grown are shown in Tables 2 and 3.

Table. Main Cultivars Grown in Japan (table grapes, 1997)

Cultivars	Area (ha)	Proportion (%)
Kyoho	6,531	33.0
Delaware	4,878	25.0
Campbell Early	1,500	7.6
Pione	1,400	7.5
Muscat Baily A	864	4.4
Niagara	492	2.5
Kosyu	470	2.4
New Baily A	382	1.9

Table. Main Cultivars Grown in Japan (for Wine and Juice, 1997)

Cultivars	Area (ha)	Proportion (%)
Yamabudo (*V. coignetiae*)	124.2	19.8
Muller Thurgau	102.7	16.4
Chardonnay	96.3	15.3
Merlot	88.6	14.1
Cabernet Sauvignon	69.9	11.1
Seibel	69.7	11.1

Some new cultivars for table grapes such as 'Aki queen,' 'Honey Seedless,' 'Honey Black,' 'North Red,' and 'Sunny Rouge,' have been bred at the National Institute of Fruit Tree Science.

Table. Description of Table Grape Cultivars Bred in Japan

Cultivar (Polyploidy)	Parentage	Fruit Description
Aki Queen (4x)	Kyoho x Kyoho	End Aug., very large, bright red, excellent quality
Fujiminori (4x)	Ikawa 682 x Pione	End Aug., very large, dark purple-red, good quality
Honey Seedless (3x)	Kyoho x Concord Seedless	Mid-Aug., medium berry (GA), greenish yellow, seedless
Kaiji (2x)	Flame Tokay x Neo Muscat	Early Oct., large, bright red, excellent quality
King Dela (3x)	Red Pearl x Muscat of Alexandria	Early Aug, small (GA), red-brown, seedless, good quality
Koshu (2x)	Unknown	Early Oct., medium, red-purple, medium quality
Kyoho (4x)	Ishihara Wase x Centennial	End Aug., very large, violet-black, excellent quality
Mirei (3x)	Red Queen x Koshu Sanjaku	Mid-Aug., medium (GA), yellow-green, good quality
Muscat Bailey A (2x)	Bailey x Muscat Hamburg	Mid-Sept., very large, bluish black, medium quality
Neo Muscat (2x)	Muscat of Alex. x Koshu Sanjaku	Early Sept., large, greenish yellow, muscat flavour
North Red (2x)	Seneca x Campbell Early	Mid-Aug., medium, red-brown, sweet, foxy
Pione (4x)	Kyoto x Cannon Hall Muscat	Early Sept., very large, violet-black, excellent quality
Rosario Bianco (2x)	Rosaki x Muscat of Alexandria	Mid-Sept., very large, greenish yellow, excellent quality
Sekirei (2x)	Early sport of Kaiji	Mid-Sept., large, bright red, excellent quality
Suihou (4x)	Pione x Centennial	Early Sept., very large, green-yellow, excellent quality
SummerBlack (3x)	Kyoto x Thompson Seedless	Early Aug., large (GA), purple-black, seedless, good quality

PRODUCTION OF PLANTING MATERIAL

Private nurseries produce most of the planting material such as rootstocks from cuttings, and then scions of commercial cultivars are grafted. Saddle grafting has produced over 90 percent of grape planting materials. Another popular grafting method is cleft grafting.

In the early 1920's, '3309,' '3306,' and '101-14' rootstocks were introduced to Japan. After that, the '420A' and 'Teleki' strains were also introduced. Growers have chosen 'Teleki 8B,' '5BB,' and '5C' because of their excellent characteristics, including high berry quality, early ripening, and wide soil adaptability. These are the reasons why they are still the major rootstock varieties in Japan.

ESTABLISHMENT OF VINEYARDS

Land Preparation

The land selected for vineyards should have a gradient of less than 5 degrees, underground watertable below 1 m, long hours of solar radiation, average annual rainfall of 1,000 mm, and, in particular, a low amount of rainfall during the harvest season. Long drainage pipes with small holes should be buried to drain underground water and provide aeration in the soil. In sloping land, prevention of surface soil erosion is very important, and, therefore, it is necessary to cover the surface with a species of grass such as Italian ryegrass.

Planting Season

There are two planting seasons, in spring and in autumn. Spring planting is done at the beginning of March, before commencement of active growth. Autumn planting, carried out between the beginning of October and the end of November, is superior to spring planting for better growth of vines.

Spacing

It is important to enlarge the canopy after planting using a horizontal trellis method of support for training the vines. The canopy enlargement depends on the kinds of cultivars, pruning methods, soil fertility, planting material quality (virus-free), cultivation style (such as root-restriction cultivation), and rootstocks. The standard planting density for 'Delaware' used to be between 80 and 240/ha, but proper spacing has recently been examined under different conditions.

CARE AND MANAGEMENT OF VINEYARDS

Training and Pruning of Plants

Most vineyards are furnished with a horizontal trellis. The horizontal trellis is very popular since most of the grape-growing areas are hit by typhoons almost every year. A horizontal trellis is more effective in minimizing wind damage. The rainy season through June and early July coincides with the period of berry enlargement or ripening in open cultures. The shortage of sunshine hours and the humid conditions tend to retard berry development and promote fungal diseases.

Leaves on a horizontal trellis can utilize the sunshine more efficiently and minimize the spread of diseases. Shoot growth also tends to be too vigorous to ensure good berry set and berry growth during the warm and wet rainy season. Enlargement of the canopy to control shoot growth can be easily performed in this training system.

Grapevine management, such as cluster trimming, GA treatment, and berry thinning, which are indispensable in order to produce high quality grape bunches, is easily conducted under a horizontal trellis. Recently, a flexible horizontal trellis that allows adjustment of the trellis angle to the stature of the grower has been developed to save labour hours and enhance management efficiency under protected culture.

There are two principal pruning methods, namely, short cane pruning (severe spur-pruning) and long cane pruning. The objectives of short cane pruning are to leave only 2 to 3 buds on one lateral shoot and maintain straight primary shoots. This pruning method has been mainly used in the western part of Japan. This kind of heavy pruning is easy for farmers to carry out without using a special technique. On the other hand, one year old canes that elongated in the previous year are pruned leaving several buds in the long cane pruning method. This pruning system has been used in the eastern areas of Japan.

Application of manure and fertilizers

Standard amounts of manure and fertilizers used in the Yamanashi area. The amount of fertilizer depends on the soil type, the age of plants and the cultivar. Young vines, virus-free vines, and tetraploid cultivars should receive less nitrogen to suppress excess growth because they exhibit very vigorous cane growth. Most fertilizer is supplied as basal fertilizer between October and the beginning of November.

Table. Standard Amounts of Fertilizers Used in Some Major Cultivars

Cultivars	Yield (t/ha)	Amount of Fertilizer(kg/ha)				Amount of Manure (t/ha)
		N	P_2O_5	K_2O	CaO, MgO	
Delaware	15	150-170	170-180	150	800	15
Kyoho, etc.(tetraploidy)	12-14	60	150	120	1000-1500	15
Neo MuscatKaiji	18-22	160-200	150-170	150-160	1500	20
Muscat Bailey A Kosyu	22-24	130-210	190-260	130-150	1000-2000	15

Weeding

There are three different kinds of surface soil management. One is sod culture, particularly in sloping land, because grass prevents surface soil erosion during the rainy season. Grass supplies organic matter into the soil, and grass roots also improve the physical conditions in the soil. However, grapevines and grass sometimes compete aggressively for water and nutrients under conditions of drought stress in shallow soils.

Another management method is known as clean culture, mainly used in flat vineyards. Herbicides such as 'glyphosate (isopropylamine)' and 'diquat paraquat' are sprayed in spring and summer. However, the soil surface surrounding the main cane is covered with straw to prevent vines from drought stress and nutrient competition and which also supply organic matter.

Irrigation

Grapevines require the most amount of water during two periods of growth, usually before germination and during berry enlargement. On the contrary, water is restricted to improve berry quality during the maturation period. The number of irrigation's depends on the soil type and the growth stage of the crop. About 30 mm of water per irrigation is sufficient during the summer season for a vineyard on clayey soil. Vineyards on clayey soils are watered at longer intervals than those on sandy soils. Soil moisture should be recorded with a pF meter and kept between pF2.3 and 2.5. Use of sprinklers is the most common irrigation system in Japan. In addition to irrigation, it is possible to use the sprinklers to spray some chemicals on a flat canopy developed on a horizontal trellis. Drip-irrigation and misting nozzles are also used under protected culture.

Pests and Diseases

a) Pests: The main pests of grapes in Japan include yellow tea thrips, grape borer, grape clearwing moth, grape phylloxera, grape leafhopper, swift moth, mealy bugs and mites. Thrips and borers are notorious pests in Japan. In warmer areas, the occurrence of yellow tea thrips is 7 to 8 times per growing season, but it is 5 to 6 times in cooler areas of Japan. Thrips over-winter under grape bark, dropped leaves, or in soil. Timely sprays are given before blooming, after flower dropping, and during the maximum occurrence period, from mid-July to early-September. Synthesized pyrethroid can be sprayed to effectively control this pest.
b) Diseases: *Vitis vinifera* is subject to numerous diseases. Some of them cause serious problems in all grape-growing areas, while some others are of local importance. Among fungal diseases of leaves and fruit, powdery mildew, downy mildew, anthracnose, grey mold, and ripe rot are prevalent. Crown goal is an important bacterial disease.

Leafroll is caused by viruses GLRaV-1 and GLRaV-3 which have been detected with ELISA. It is reported that both of them propagate and spread through grafting. In addition, GLRaV-3 is transmitted through comstock mealy bug (*Pseudococcus comstcki*), and GLRaV-1 has shown evidence of sap transmission. In Japan, several cultivars show latent infection by corky bark, a kind of rugose wood complex. Although damage is not serious, grapevine berry inner necrosis caused by GINV (grapevine berry inner necrosis virus) was found in Ibaragi Prefecture in 1984. This virus shows graft and sap transmission. Yellow speckles caused by viroids have not been detected in Japan so far. Many cultivars show lateral infection with Fleck that is thought to cause Aji-nashika (low-sugar berry) disease.

Physiological Disorders

a) Shatter (floret dropping): This disorder is a phenomenon characterized by florets dropping after blooming. Lack of proper fertilization, bad nutrient conditions, and physiological stress (water deficit or high temperature) cause florets to drop. Several cultivars, such as 'Kyoho,' 'Pione,' and 'Muscat of Alexandria,' are susceptible to this type of disorder.
b) Sunscald: Sunscald has been mainly observed on 'Muscat of Alexandria' under glasshouse conditions from the late first stage to the late second stage of berry growth. There are different kinds of disorders on the berry surface, including blotch, berry shrink, and sunburn. These disorders are commonly caused by denaturation of the vascular bundle. The sunscald phenomenon might be related to 'water berry' or 'stem necrosis,' which are also caused by a disorder of the vascular bundle. Therefore, the mechanism of this disorder needs to be clarified as soon as possible.
c) Berry Cracking: Cracking during maturation depends on cultivars and cultivation conditions. 'Delaware,' 'Olympia,' and 'Red queen' crack easily. Berry cracking is observed in diluvial soil more frequently than in alluvial or volcanic ash soils. Research on grape skin indicates that cracking is due to inferior texture of the skin, the long oval shape of the berry with a pericarp that can be easily distorted, and a thinner cell wall of the sub-epidermal cell layer. It is important to avoid wet soil conditions after long periods of water deficit during the growth stage to protect berries from cracking.

In addition to these disorders of the berry, low berry colouration is a problem caused by high temperature during the maturation stage, heavy fruit load, bad nutrient conditions, virus infections, or low light intensity, among others.

d) Less Growth due to Cold Damage: This is a kind of cane growth disorder. Less sprouting and slow growth of canes are observed in the spring, particularly in cold areas, due to cold damage in association with water deficit. It is therefore necessary to improve freezing resistance of both cultivars and rootstocks. New cultivars with high freezing tolerance bred by the National Institute of Fruit Tree Science (NIFTS), such as 'North Red' and 'North Black,' have been introduced in northern areas of Japan.

Grape Quality Improvement and New Cultivation Systems

a) Inducing Seedless Grapes by GA Treatment: Consumers prefer seedless grapes because they are easy to eat. Success in producing seedless 'Delaware' grapes with GA application remarkably increased the consumption of this kind of grape. Fruit clusters are dipped with 100 ppm of GA solution 2 weeks before blooming and again after blooming. The pre-bloom treatment makes the berries seedless, while the post-bloom application stimulates berry enlargement. These treatments cost about 300,000 yen (US$ 2,800) for the chemicals and require 300 to 400 hours of labour per hectare. Seedless 'Muscat Bailey A' grapes have also been successfully produced by the same treatment. For 'Kyoho' and 'Pione,' a modified treatment with lower concentration of GA and a later application time has been developed.
b) Prolonged Period of Fruit Supply by Protected Cultivation and New Storage Systems: Fruit harvesting under an open culture begins in late July or early August for early cultivars, followed by mid-season cultivars in late August and October, and ending in late December for late cultivars. Harvest time is markedly advanced in heated houses and grapes marketed earlier are sold at a higher price. Application of heating can begin as early as late November or December. In such cases, fruit is harvested from April to May. In vines heated in late January or February, fruit is harvested from May to July or August.

Cold storage of grape bunches can delay the time of shipping. 'Kyoho' bunches can have their marketable period extended by 30-50 days when wrapped with plastic films and stored at 0° C and 80 percent relative humidity. New storage methods with minus ion and ozone that were developed by NIFTS can tremendously extend the storage period by over 70 days.

Although heated cultivation and new storage systems can extend fruit supply from early April to December, it is difficult to supply grapes during the spring season from late January to March. For this reason, the suppression culture, which requires the use of UV-less plastic films, and the double-cropping culture have been examined.

c) Double-cropping: Research on the double-cropping of 'Muscat of Alexandria' under protected cultures began in the 1960's in Japan. Grape growers have been interested in double-cropping to enhance yield and income. The first cropping is in late June or early July, and the second one is in December or January, when high prices are expected. In this cultivation, specific techniques such as temperature control, breaking of dormancy, pruning, long-day treatment by fluorescent lamps, and enhanced CO_2 application are very important. Some growers in Okayama and Shimane prefectures have adopted the double-cropping system.

d) Root-restricted Culture and Soil-less Culture: In addition to developing the protected cultivation techniques, root-restricted culture and box culture have been examined. These culture systems promote good control of vine vigour, stable production, and fruit quality improvement.

In a soil-less culture, a hydro-culture or artificial solid materials such as a rock wool system are used to produce grapes. This culture has been tested to control grape production automatically.

Use of Plant-Growth Regulators in the Vineyard

The use of plant-growth regulators in vineyards to control growth and fruit quality has been studied for many years. Several plant regulators have been known to improve grape production in Japan. Gibberellic acid (GA) has been used since the 1950's to grow seedless grapes and increase berry size. The effect of GA on the promotion of berry maturation and bunch elongation has been observed. Forchlorfenuron (1-(2-chloro-4-pyridyl)-3-phenylurea) liquid has also been used to increase berry size and prevent floret dropping (shatter).

Streptomycin is also effective for seedless grape growth and promotion of berry maturation. To prevent florets from dropping, BA (6-N-benzylamino purine) has been used. Calcium carbonate has been employed to reduce the damage by water-soluble copper powder. Forchlorfenuron and BA, ethephon (2-chloroethylphosphnic acid), and mepiquatchloride (1,4-dimetyl piperidinium chloride) are effective to prevent florets from dropping. C-MH (maleic hydrazide choline) has also been used to suppress excess vegetative growth of the cane.

HARVESTING AND YIELDS

One of the indicators of the harvest season for table grapes in 'Kyoho' is a sugar:acid ratio of 30 to 40 in late August. Fruit quality, including bunch appearance and skin colour, is quite important for shipping in Japan. Using the standard colour chart, skin colour can be determined. Yields depend on cultivars and culture type. An open culture of 'Kyoho' has an average yield of 12 to 15 metric tonnes per hectare. The average sugar content (Brix) is 17-18, and the content of tartaric acid is 0.8-0.9. 'Delaware,' 'Campbell early,' 'Muscat Bailey A,' and 'Neo Muscat' yield 15 to 18, 24, 23, and 20 to 24 t/ha of table grapes, respectively. Proper yields per hectare are decided by average cane length and the number of new cane per/m^2 based on the material production theory through photosynthetic capacity. Heavy fruit load causes low sugar content and a colouration that is not acceptable for table grapes. Moreover, grapes with berry skin disorders do not sell or must be sold at a low price. Proper yields promise high fruit quality and income.

MARKETING

In 1997, total grape production was 250,900 metric tonnes, 87.4, 11.0, 1.4, and 0.2 being the percentages for the production of table grapes, wine, juice, and canned fruit, respectively. Only 14 tonnes of table grapes were exported, while 7,649 tonnes were imported in 1997. The grapes marketed earlier were sold at a higher price.

POTENTIAL FOR GRAPE-PRODUCTION DEVELOPMENT

Grape production in Japan has not changed dramatically during the past 20 years. However, wine consumption is gradually increasing these days because of its functional components, such as polyphenol, which are thought to be effective in suppressing human senescence.

Therefore, high quality wine grape production might develop more in the future. Table grapes must always be of higher quality because higher quality fruit can be sold at a higher price. Stable production of high quality table grapes brings more income to farmers. The potential exists for the development of exports to overseas countries, since only 12 metric tonnes of grapes were exported in 1997.

CONSTRAINTS IN GRAPE PRODUCTION DEVELOPMENT

Although wine consumption has increased in recent years, the total consumption of not only grapes but also all kinds of fruit has been continuously on the decline in the last two decades. Grape farmers who have small grape farms cannot increase their vineyard area or their production to improve their income further. This consumption decrease is one of the biggest constraints in Japan. A second problem is that, as they age, most farmers suffer the consequences of the long, hard, and intensive labour required for berry thinning and bunch trimming, among other tasks that are necessary to produce high quality grapes. The requirement of intensive labour is the reason why there are few young successors. Therefore, new techniques to achieve low-cost production and save labour are required to solve those problems.

GOVERNMENT POLICIES AND PLANS FOR GRAPE RESEARCH AND DEVELOPMENT

Breeding Objectives

The major breeding objectives are to select excellent cultivars adapted to Japanese climatic conditions. Seedlessness, large berry size, good appearance, and high quality are the principal considerations for table grapes. The use of triploid or tetraploid cultivars is advantageous.

For wine grapes, to improve the wine quality of old Japanese cultivars, back crossing of *V. vinifera* grape to these cultivars is being carried out as the demand for wine increases.

The selections are propagated and distributed to several research stations with different climatic and soil conditions for multi-location testing. Productivity, fruit quality and wine quality are tested to select promising cultivars for both table and wine grapes.

Disease resistance is another important objective. Most grape cultivars are susceptible to many diseases. Inoculation techniques have been evaluated in downy mildew, anthracnose, ripe rot, and grey mold in Japan.

Biotechnology

The development of biotechnology is very important to improve grape cultivars. Grapes are heterozygous plants, and genetic analysis for important agronomic traits has not made appreciable progress. Haploid plants are of considerable value for genetic analysis and crop improvement. Protoplast fusion and somatic hybridization enable the achievement of wide hybridization and partial gene transfer.

Although cross-hybridization is generally used for plant breeding, it results in plants inheriting both favourable and unfavourable characters at the same time. On the other hand, only the target gene can be transferred to the plant by means of genetic engineering. Gene-delivery systems such as the *Agrobacterium*-mediated system, electroporation, and DNA-coated microprojectiles can be used for this purpose. The analysis of DNA, such as RAPD, RFLP, and AFLP, is an important technique for the classification, identification, and development of molecular markers and the isolation of useful genes in grape breeding.

Physiological Aspects

A stable annual production of high quality fruit under different cultivation systems is the most important goal for both table and wine grapes. It is important to research stable berry set, control maturation, and skin colour based on plant hormone movement, water relations, photo-assimilate translocation, and gene analysis in grapevines.

Low-cost Production and New Technology for Labour Saving Culture

To achieve high quality table grapes, berry thinning and cluster trimming are the most labour-demanding management tasks. It is necessary to reduce the amount of labour and improve the efficiency of labour productivity to produce grapes at a lower cost. However, it is quite difficult to do berry thinning and cluster trimming automatically or mechanically, so breeders have been trying to breed a new cultivar that does not require berry thinning or cluster trimming.

Environmental Concerns

Fewer chemicals and less nutrients should be supplied to orchards to preserve the environment. A biological protection system using natural enemies, for example, will be researched to prevent grapevines from diseases and pests. The amount of nitrogen fertilizer used should also be examined and reviewed again from the point of view of environment preservation and plant nutrition effect.

GRAPE PRODUCTION IN AUSTRALIA

The first grapevine planting material arrived in Australia with white settlement in 1788. Today grapes are grown commercially in all States and Territories. Grape growing (wine, raisin, and table) is the largest fruit industry in Australia with production in a wide range of environments from temperate to tropical. Wine grape production and wine making is the largest and most predominant of the three viticulture industries.

There has been considerable change in production in all three industries over the last 4 years, up to 1999. Wine grape production has almost doubled to 1,076,207 tonnes and dried grape production has fallen to 119,438 tonnes, largely in response to export markets plus increasing imports of dried grapes. Table grape production has increased by almost 60 percent to 69,891 tonnes fuelled by industry expansion in sub-tropical regions and increasing exports to Asia. A feature of recent industry expansion has been the increased planting of varieties specific to the end use (wine, dried, table).

Nevertheless, the multipurpose varieties such as Sultana, Muscat Gordo Blanco and Waltham Cross continue to account for 40 percent of white grape production in Australia. Grape growers redirect their fruit from drying to processing for wine or production of table grapes as the fortunes of each industry waxes and wanes.

A comparison with other grape producing countries throughout the world shows that Australia was the 14th largest producer of grapes (tonnes) in 1997, when Australia was ranked 9th in the world for volume of wine produced. While 28 percent of production (170.6 ML) was exported, this amounted to only a very small percentage of world wine exports. Australia is a relatively small producer and exporter of table grapes and raisins by world standards.

The viticulture industries in Australia are well developed, innovative and utilize the latest production practices. Australia is ranked among the top 10 countries in terms of average grape yield (t/ha). There is a high degree of mechanization in the vineyard, particularly in the wine grape and raisin industries. A substantial research, development and extension effort continues to generate practices that improve production efficiency and grape quality.

CURRENT STATUS OF GRAPE PRODUCTION

Grapes are grown in all States of Australia but most of the production is in the temperate zone. The three largest States by production are South Australia (Riverland), Victoria (Sunraysia) and New South Wales (Riverina). Wine grapes are grown in all States, table grapes in all States except Tasmania and raisins in Victoria, New South Wales, South Australia and Western Australia.

Over the last five years wine grape production has almost doubled, raisin production declined slightly and table grape production increased moderately. Further substantial increases in wine grape production will occur during the next two to three years as non-bearing vines come into production and reach mature cropping levels.

Most of the growth in the table grape industry has occurred in sub-tropical areas across northern Australia. Production volumes will increase over the next two years when these new vineyards come into bearing. New plantings of wine grapes are expected to slow down, while raisin grapes will remain static and table grapes show a slight increase.

Table. Production of Wine grapes, Raisins and Table grapes in Australia, 1998-99 (in tonnes).

	Wine making	Drying	Table grape	Total
New South Wales	270,236	19,137	14,128	303,501
Victoria	277,869	96,788	42,391	417,048
Queensland	1,264	-	5,586	6,850
South Australia	491,621	2,764	2,149	496,534
Western Australia	32,067	749	3,531	36,347
Tasmania	3,121	-	-	3,121
Australia	1,076,207	119,438	69,891	1,265,536

A large number of varieties are grown in Australia. The area planted with each variety and tonnage utilized for wine making, drying or as table grapes. The major table grape varieties are Thompson Seedless, Red Globe, Flame Seedless and Menindee Seedless. Menindee Seedless is the predominant variety of plantings in northern Australia. Other table grape varieties grown include Cardinal, Emperor, Ribier (Alphonse Lavelle), Marroo Seedless, Calmeria, Ohanez, Purple Cornichon and Waltham Cross.

Table. Area under Grape Varieties and Production Extents (1998-99).

Variety	Area of Vines (ha)			Grape Production (tonnes)			
	Bearing	Non-bearing	Total	Wine making	Drying	Table grape	Total
Red Grapes:							
Barbera	111	105	216	669	-	-	669
Cabernet Franc	559	123	687	5 365	-	-	5 365
Cabernet Sauvignon	13 629	7 541	21 169	127 494	-	193	127 687
Currant	889	49	938	5 515	7 189	10	12 714
Grenache	2 025	231	2 255	24 196	53	33	24 281
Malbec	399	49	448	2 857	-	-	2 857
Mataro	683	183	866	9 217	-	69	9 286
Merlot	3 465	292	6 387	31 801	16	246	32 063
Muscat a Petit Grains Rouge/Rose	296	48	345	1 411	46	408	1 865
Petit Verdot	110	261	370	1 045	-	-	1 045
Pinot Noir	2 226	770	2 996	19 668	26	267	19 960
Ruby Cabernet	1 102	876	1 978	18 414	-	49	18 462
Sangiovese	311	129	440	403	-	-	403
Shiraz	16 944	8651	25 596	192 330	36	477	192 843
Tarrango	87	20	107	2 199	-	-	2 199
Other red grapes	2 528	601	3 129	7 228	658	20 179	28 065
Total Red Grapes	45 363	22 563	67 925	449 809	8 025	21 929	479 762
White Grapes:							
Chardonnay	15 298	1 558	16 855	210 770	-	299	211 069
Chenin Blanc	884	55	939	16 621	-	-	16 621
Colombard	1 382	266	1 648	34 781	-	-	34 781
Crouchen	83	5	87	1 136	-	-	1 136
Doradillo	306	8	314	6 597	-	1	6 598
Marsanne	137	63	201	1 878	-	-	1 878
Muscadelle	230	17	247	2 106	-	7	2 113
Muscat a Petit Grains Blanc	261	8	269	3 334	5	24	3 363
Muscat Gordo Blanco	2 924	81	3 005	58 017	2 575	112	60 703
Palomino	161	2	163	2 581	-	-	2 581
Pedro Ximenes	146	-	146	1 981	-	-	1 981
Riesling	3 190	157	3 347	30 144	-	-	30 144
Sauvignon Blanc	1 926	487	2 413	22 834	-	100	22 934
Semillon	5 307	737	6 044	80 191	-	-	80 191
Sultana	12 943	696	13 639	117 783	105 982	26 339	250 103
Taminga	44	-	44	862	-	-	862
Traminer	531	7	538	5 357	-	-	5 357
Trebbiano	689	1	690	10 482	-	-	10 482
Verdelho	708	267	975	7 290	-	-	7 290
Viognier	79	51	130	494	-	-	494

PRODUCTION OF PLANTING MATERIAL

Production, supply and maintenance of healthy planting material are well organized and wholly undertaken by industry. There are three tiers involved, namely, the national, state and individual nursery operator levels. The Australian Vine Improvement Association (AVIA) is the national organization responsible for developing standards in vine health maintenance and product quality. Each state has a vine improvement organization (VIO) whose role is to produce healthy cuttings as specified by AVIA, for sale to nurseries.

Privately owned nurseries propagate vines for sale to grapegrowers, preferably using cuttings from VIOs but also from commercial vineyards in times of short supply.

Table. Rootstock Cuttings Sold in Australia, 1998.

Variety	Number Cuttings
Ramsey	2,902,841
Schwarzmann	1,863,870
140 Ruggeri	908,332
1103 Paulsen	779,218
K51-40	543,471
99 Richter	536,900
5BB Kober	522,775
101-14	449,059
5BB Kober 5A	429,735
5C Teleki	326,219
SO4	296,466
K51-32	107,855
110 Richter	89,059
Dog Ridge	87,645
Teleki C	36,750
Freedom	29,164
Harmony	22,600
J17-48	18,150
420A	11,210
J17-69	2,200
Rupestris du lot	1,836
34 EM Foex	1,150
Riparia Gloire	400
St George 15-74-2165	215
ARG 1	100
1202 Couderc	100
5A Teleki	20
1616 Couderc	10
Total	**9,967350**

In recent years a substantial effort has been made at all levels of the industry to improve the health status of grapevine planting material in Australia. Virus indexing of government germplasm collections using more sensitive techniques is almost complete and indexing of the VIOs source area vineyards, from which cuttings are produced and sold to nurseries, is well advanced. These vines will now be indexed every 2 to 3 years.

AVIA and the state VIOs have jointly developed sanitation protocols for use by nurseries, in particular a hot water treatment (50°C, 30mins) of cuttings for the control of crown gall and phytoplasmas. A quality assurance scheme has been developed by AVIA, which governs the health status and quality of planting material sold by participating nurseries.

The types of planting material available are dormant cuttings, vines on own roots and vines grafted to rootstocks. Most planting material is purchased from nurseries and planted as vines with roots. Some growers continue to plant callused cuttings directly into the vineyard but this practice is dying out. Bench grafting using machines built for the purpose is the most widely used technique by nurseries. A small number of nurseries continue to graft in field nurseries. One specialized nursery has recently started green micro-grafting which enables the production of bench grafted vines all year round.

Throughout the 1990's there was a large increase in the number of vines grafted to rootstocks. More than 20 rootstocks are used in the Australian viticulture industries, primarily for salt tolerance, increased productivity and nematode tolerance. Rapid expansion of the grape industries, mostly wine grape and table grape, has seen large increases in the number of vines produced each year but demand exceeded supply which led to shortages of propagating material. It was estimated that in some years, up to 60 percent of the propagating material used by nurseries came from commercial vineyards and hence had a doubtful health status and quality.

ESTABLISHING VINEYARDS

Site Assessment

Long-term profitability and ease of management is greatly enhanced by detailed assessment of the proposed vineyard site. Firstly, the optimum district for producing the desired fruit (wine) quality or (table grape) harvest time is determined using long-term weather data. Once a vineyard site is chosen a detailed assessment of water and soil resources is undertaken.

The water survey will include assessment of on-farm water resources, access to off-farm water supplies and quality of the water. A water sample is then collected and analyzed. Grapevines are not highly sensitive to salt but water supplies with a total salinity exceeding 2000 microSiemens per centimetre, sodium content greater than 20 milliequivalents per litre or chloride content greater than 4 meq/L should be avoided. On-farm water resources are surveyed to determine the quantity of water available from dams or underground sources through bores, as well as the catchment area so that recharge rates can be calculated. Much of Australia's viticulture is located in irrigable areas where water supply is reliable and well managed. Many of the new vineyard developments rely upon pumping water from rivers and streams, and this requires a licence issued by a statutory organization. The quantity and timing of rainfall in relation to the growing season, which varies considerably throughout Australia, is included in the assessment of water resources. The water survey is the major determining factor in the design of the vineyard irrigation systems.

Soil surveys are increasingly used in vineyard planning. They are the major determinant of vineyard layout and indicate what soil ameliorants and physical modifications are required before planting.

A simple soil survey involves collecting soil samples for chemical analysis and nematode assay, whilst a detailed survey involves digging soil pits in a 100m grid across the vineyard site. Soil pits are used to determine variability across the site, presence of compaction or impermeable layers, and subsoil chemical status.

Soil Preparation

Soil preparation involves applications of lime or dolomite to adjust pH; superphosphate, as most Australian soils are low in phosphorous; and applications of magnesium, zinc and boron where these nutrients are locally deficient. Ripping along the vine row and sometimes of the entire site is used to break up compaction and impermeable layers and increase soil depth. Where nematodes are present tolerant rootstocks are chosen and ordered from commercial nurseries.

Planting

Immediately prior to planting the soil is cultivated. Planting is commonly carried out in late winter-early spring using dormant vines. Numerous mechanical devices are available to assist with planting of vines. The recent introduction of container grown vines has enabled the planting season to continue through to mid-summer (Dec/Jan), even in sub-tropical areas.

Vine Spacing

Vine spacing varies slightly with grape type (wine, raisin, table), age of vineyard and district. Rows are mostly 3.0-3.6m wide with vines 1.5-2.4m apart. Spacings are commonly at the higher end of the range in more recent vineyards because of larger trellis designs, plus rootstocks and improved management practices contributing to greater vine vigour. A few medium density (2.0m x 1.0-1.5m) wine grape vineyards have been planted but these spacings are not suited to the climate of most grape production districts in Australia.

The use of 'vine tubes' to promote faster growth and establishment of young vines has become widespread over the last 5 years. A variety of tubes are available, differing in design, colour, length and material. Vine tubes are rarely used and considered unnecessary in northern Australia, as vines establish quickly and carry their first crop 18 months after planting.

VINEYARD MANAGEMENT PRACTICES

Trellis Systems

The most common trellis used for table grapes is a 6 or 8-wire large Y shaped trellis. The 6-wire Sloping T trellis is far less popular. Both trellis designs have two cordon wires and vines are trained to a quadrilateral. Moveable foliage wires are not used nor hand positioning of shoots. Both trellis designs provide good separation of fruit and foliage. The Y trellis is preferred as it provides unobstructed access from both sides for harvesting and pre-harvest hand manipulation of bunches. The application of plastic vine covers is a little easier with the Sloping T trellis.

Pruning

Most varieties grown are spur pruned even in the tropical areas (Mareeba). Thompson Seedless and Menindee Seedless require cane pruning and there has been a recent shift to cane pruning of Red Globe for greater fruit yield. All pruning of table grapes is done by hand. A large proportion of the wine grape area planted throughout Australia is machine pruned or not pruned at all.

Hydrogen cyanamide (Dormex ®) is applied after pruning to enhance percent bud burst, promote uniform bud burst and advance bud burst time in the early season districts. The interval between application and bud burst increases with latitude from 10 days (Mareeba) to 25 days (St George). Notwithstanding the use of hydrogen cyanamide bud burst is more protracted and the percent bud burst is less than in temperate areas. Our understanding of vine dormancy, interaction with the environment and the manipulation of bud burst needs to be improved if vineyard productivity in sub-tropical areas is to be increased.

Vine Nutrition

Inorganic fertilizers are predominantly used for maintaining vine nutrition. Annual dry matter analysis of petiole samples is commonly used for determining the nutrient requirements of grapevines. Some growers also use soil analyses but generally only every second or third year.

Most grape growers apply inorganic fertilizers through the irrigation system (fertigation) because it is easy, provides a high degree of flexibility and control, minimizes leaching losses in summer rainfall districts and ensures the nutrients are available when required by the plant. Broadcasting of fertilizer for incorporation by rain is uncommon and generally used only for particular fertilizer products.

Annual fertilizer application rates for table grapes in the sub-tropics range from 20 to 40 kg/ha N, 10 to 20 kg/ha P, and 60 to 100 kg/ha K. Dolomite or lime is regularly applied to adjust soil pH. Foliar fertilizers are increasingly being used for the application for micro-nutrients (Zn, B, Mg, Mn).

Only a small number of growers use organic fertilizers but the interest in organic fertilizers is increasing sharply. Organic fertilizers are invariably expensive per kilogram of nutrient, not easily accessible and often in short supply.

Vineyard Floor Management

Standard practice is a weed free strip along the vine row and a permanent plant cover between rows. This minimizes soil erosion and facilitates vehicle access after rain, when the timely application of fungicides for disease control is important. The inter-row plant cover is usually self sown, although in vineyards in temperate southern Australia there is a shift towards the planting of specialist species. The inter-row plant cover is regularly slashed and the under-vine strip maintained weed free with herbicides. Cultivation is generally not practiced.

Irrigation

Irrigation is essential for profitable table grape production. Drip irrigation is the standard technique used. Furrow irrigation or under-vine sprinklers are not used because of limited water supplies and both methods of irrigation are found only in irrigation districts in southern Australia.

Scheduling of irrigation is increasingly used over the traditional practice of regular application of a prescribed amount. Scheduling methods vary from rather inexpensive and simple (crop factors, evaporation and rainfall data, tensiometers, gypsum blocks) to highly technical and more expensive techniques (neutron probes, capacitance probes). Irrigation scheduling provides flexibility of management for growers, leads to efficient use of water, avoids leaching, and maximizes vine performance.

Pest Management

Low chemical input is a characteristic of Australian viticulture and over the last 10 years there has been a shift to greater use of biological control agents. The viticulture industries are moving from routine to strategic spray programmes and this is most advanced in the wine grape industry where higher fruit damage levels are acceptable.

The success of strategic spray programmes is the accurate prediction of pest outbreaks and disease infection periods. This requires monitoring in the vineyard for the presence and development stage of a pest and recording weather data. The 'AusVit' software package analyzes the weather data and vineyard observations to produce recommendations which assists grapegrowers to make spray management decisions.

The list of insects is very similar throughout the grape production regions of Australia. The one notable exception is the occurrence of Queensland Fruit Fly in northern Australia only, which is a major pest of table grapes in this region, and its absence in southern areas. A biological insecticide *Bacillus thuringiensis* is available for control of light brown apple moth (LBAM) and grapevine moth caterpillar. The predatory wasp *Trichogramma* is also effective against LBAM. The pest mites (blister, bunch, rust) are controlled organically by applications of wettable sulphur and several predatory mite species endemic in Australian vineyards. Growers maintain predatory mite populations by the careful selection of safe pesticides. Queensland fruit fly and cadydids (grasshoppers) are the two significant pests of table grapes in northern Australia and require regular control to prevent serious damage.

Diseases are of far greater concern and impact on grape production throughout Australia. Downy mildew (*Plasmopara viticola*), Botrytis (*Botrytis cinerea*) and anthracnose (*Elsinoe ampelina*) are the most damaging diseases of table grapes in northern Australia. Under the high summer rainfall conditions common throughout northern Australia the damage to fruit and foliage can be extensive and severe.

The standard spray programme for table grape production involves regular applications of protectant fungicides with the timing determined by growth stage, weather forecasts and disease incidence the previous season. Copper oxychloride and mancozeb are the major fungicides used for downy mildew control. The eradicant fungicides metalaxyl, phosphorous acid, benalaxyl and oxadixyl are important for the post-infection control of downy mildew, particularly in Queensland where storms and two to three days continuous rainfall are common during the growing season.

The spray programme for Botrytis is very specific and consists of two sprays during flowering, one spray before bunch closure, and one to three sprays during pre-harvest. To avoid resistance developing, particular attention is paid to using a fungicide from a different chemical group at alternate applications. Resistance to benomyl and procymidone exists in some grape districts, particularly Western Australia where resistance is extensive throughout the table grape industry. The biological fungicide*Trichodex* is available but is less effective under high disease pressures.

Control of anthracnose is possible only with regular applications of the pre-infection dithiocarbomate fungicides or dithianon for woolly bud and whenever new growth is present during wet weather. Once an outbreak of anthracnose has occurred it can take three to five years of good control practices to clean the disease from the vineyard. Fungicides used for anthracnose will also control Phomopsis (*Phomopsis viticola*).

Powdery mildew (*Uncinula necator*) is less important in the wet and humid climate throughout Queensland, but is the major disease of table grapes at Ti Tree (Northern Territory). A programme of three or four sprays at seven to fourteen day intervals from bud burst is very effective in preventing powdery mildew. Wettable sulphur is commonly used early in the season, then as the daytime temperatures exceed 30°C the demethylation inhibitor fungicides are used.

Table Grape Quality

Gibberellic acid is routinely used with Thompson Seedless and Flame Seedless to thin flowers in the inflorescence and increase berry size. Gibberellic acid is not applied to Menindee Seedless because it reduces fruitfulness, or Marroo Seedless because it leads to a large number of green, shot berries in the bunch at harvest.

Plastic rain covers to protect fruit from pre-harvest rain damage are a recent innovation within the table grape industry. The rain covers are spread over the canopy of the vine just after veraison (approximately four to six weeks before harvest). Their use has become widespread in the Sunraysia, is rare in Queensland and is standard practice at Ti Tree with Thompson Seedless.

Trunk or cane girdling is not used in northern Australia because the effectiveness of the technique in sub-tropical environments is not known and growers generally have no experience.

HARVESTING AND YIELDS

Table grapes are hand harvested, packed into cardboard cartons, cooled on farm and transported to market in refrigerated trucks. Most growers in northern Australia pack in the vineyard as it is cheaper, the fruit is destined for the domestic market, and is easy to manage as only one line of fruit is being packed. Shed packing is used where there is some damage requiring extra work to clean the fruit, and when the fruit is destined for an export market, multiple lines are packed simultaneously, and for varieties sensitive to handling (e.g. Thompson Seedless).

Polystyrene cartons are rarely used for table grapes since the major retail supermarket chains decided two years ago they would no longer accept produce in polystyrene, because of the waste disposal problems of this bulky material in the cities. This is despite the superior insulation properties and better protection of fruit during transport. Plastic bunch bags were introduced three years ago and their use has increased in response to demand by the retail supermarket chains. Sulphur dioxide pads are not used in northern Australia where fruit is consigned to the domestic market, but are included in cartons of table grapes exported from Sunraysia and Swan Valley. Northern Australian table grape yields average 7 kg/vine (8.75 t/ha) and only 3 kg/vine (3.75 t/ha) for the major variety Menindee Seedless. In the Sunraysia the average yield is 12 kg/vine (18.5 t/ha) and can exceed 20 kg/vine (31.0 t/ha) for Red Globe.

Low bud fruitfulness due to the growing conditions in the sub-tropics is the major reason for relatively low yields across northern Australia. Low percentage bud burst and hence fewer shoots per vine is a major factor. The short duration of each growth stage (bud burst to harvest) leads to smaller bunches and berry size, which contributes to low yields and impacts on fruit quality.

Average yield of raisin grapes is 25 t/ha (fresh weight) and 6 t/ha of dried grapes to greater than 25 t/ha for wine grapes, largely due to the production system. A sizeable and increasing percentage of the wine grape crop is harvested mechanically. This provides cost savings and is beneficial for maintaining fruit quality. In the modern raisin grape vineyards canes are cut to allow drying of the fruit on the vine and subsequent machine harvesting. Considerable development of mechanical harvesting techniques for wine grapes and raisins has occurred in Australia.

MARKETING

Table grapes are sold by two methods in Australia. Firstly, the traditional practice of consigning fruit to a wholesale agent at the central market in each of the capital cities for sale on commission. More recently, an increasing amount of fruit is being sold directly to the retail supermarket chains under contract for an agreed price.

There are no table grape exports from districts across northern Australia as this production is early season and commands a high price on the domestic market. Table grape exports are predominantly from the Sunraysia and Swan Valley. In 1998-99 table grape exports reached a record high of 31,017 tonnes valued at $69.1 million.

The major markets were Hong Kong and Singapore accounting for 40 percent and 25 percent, respectively, for a value of $44.3 million. The major varieties exported are Thompson Seedless, Red Globe, Flame Seedless and Menindee Seedless.

Table grape imports have been permitted only from New Zealand and are less than 5 tonnes each year. The Australian Quarantine and Inspection Service announced on 14 January 2000, that table grape imports from California would be permitted. Fruit is expected to arrive from July/August and continue through to December.

Export sales of Australian wine have grown dramatically from 21.3 million liters in 1986-87 to 215.5 million litres in 1998-99. The major markets are Europe, North America and New Zealand. Wine exports to countries throughout Asia have also increased significantly in recent years and this region has become the focus for further export growth.

INDUSTRY POTENTIAL

The table grape industry has grown significantly in recent years, mostly in northern Australia and the potential exists for continued expansion in this region. Water and land resources are available and support infrastructure is now well established. The potential exists to greatly increase domestic per capita consumption during the October to December period, which will in turn stimulate industry growth. Several large growers in Queensland have established links with exporters in preparation for export. Continued growth of table grape exports to Asia is likely to lead to industry growth in the Sunraysia and Swan Valley.

Expansion in the wine grape industry will slow but wine exports will continue to grow. Recent expansion of the wine grape industry has been into non-traditional production districts, including more sub-tropical parts of Queensland.

The raisin grape industry is likely to remain static under pressure from imports, loss of traditional export markets and as grape growers plant wine or table grapes in preference.

The major strengths of the Australian viticulture industries are the extensive research and development efforts and high level of skill amongst grape growers. The research and development effort has led to the current production practices and technologies in Australian viticulture to become some of the most advanced in the world. Grape growers are continually improving their skills and knowledge and adopting the technologies arising from research and development.

The geographic range of grape production districts spreads the risk of crop failure and reduces the likelihood of gaps in supply. The range of climates in which grapes are grown throughout Australia extends the supply time of fresh table grapes and enables a wide variety of wine products to be produced.

There is the potential for further advances in low chemical use production practices and the development of additional biological control methods. Scope exists for greater adoption by grape growers of currently available integrated pest management practices and low chemical input viticulture. Significant improvements in vineyard efficiency and productivity will occur as a result of continued advances in irrigation management, vine improvement and vineyard mechanization.

CONTRAINTS ON GRAPE PRODUCTION

Increasing salinity and pressure on water resources in the major grape production districts of southern Australia is a major issue for the future viability and development of all grape industries. There are huge political and environmental implications surrounding both issues. The salinity problem will be expensive and take considerable time to correct and in the meantime could result in some water and land resources becoming unusable. There are several constraints to development in the table grape industry. Firstly, the lack of new varieties which satisfy domestic and export market preferences and which are suited to local growing conditions, particularly those across northern Australia. Secondly, limited knowledge of vine physiology and performance under sub-tropical conditions is a major limitation to productivity. Specific issues include floral initiation and factors controlling bud fruitfulness, inflorescence and berry growth, vine nutrition, dormancy, the mechanism controlling bud burst, and managing vegetative growth. Major impediments to initiating and developing an export table grape trade from production districts in northern Australia are a lack of market intelligence, poor contacts in Australia and destination markets, and limited export experience amongst grape growers.

Grape growers in Australia are highly skilled, continually updating their viticulture knowledge and have a high propensity to adopting new technologies and practices. There is a comprehensive education programme providing training at all levels. Several universities train viticulturists and oenologists, the extensive network of TAFE colleges across Australia train vineyard workers in the practical skills, Government agriculture agencies provide education and training in new practices and technologies arising from research and development, and numerous secondary schools now teach viticulture. Learning also occurs in the many grape grower discussion groups organized within industry.

GOVERNMENT POLICIES AND RESEARCH AND DEVELOPMENT COMMITMENT

The Government is the major funding source for the substantial research effort within Australia on behalf of the viticulture industries. All three grape industries jointly fund research and development through levies paid by grape growers and processors (wine and raisin). The majority of the viticulture research is on wine grapes. Table grapes receive the least amount of research effort, almost all of which is undertaken in sub-tropical environments by the Queensland Department of Primary Industries, Northern Territory Department of Primary Industries, and Fisheries and Agriculture Western Australia.

Research and development is undertaken by numerous organizations throughout Australia. The major organizations at the national level are the Cooperative Research Centre for Viticulture, Commonwealth Scientific and Industrial Research Organization and Australian Wine Research Institute. The Agriculture Department in each state is another major contributor to viticulture research and development, with the level of commitment varying depending on the size of the grape industries in the state. Several universities across Australia are also involved in grape research. The current research activities of each organization can be found in their annual report or web page.

Each grape industry has a research and development corporation, which formulates research priorities, collects the levy money and then allocates it to research projects. The priorities of the wine grape industry are:

a) Grapevine variety improvement,
b) Management of pests and pathogens,
c) Grape quality,
d) Resource and vineyard management,
e) Economic prediction and assessment,
f) Technology transfer.

Research priorities of the raisin industry are available in the annual report of the Dried Fruit Research and Development Corporation. The table grape industry is not organized at the national level, so priority issues are developed at the state or local level only. The important issues in Queensland are bud fruitfulness, vine nutrition, dormancy and bud burst management, environmentally suited new varieties, quality assurance management, and integrated pest management.

Federal and State governments have policies focusing on expanding exports of food products into Asia.

This encompasses wine and table grapes. To achieve this objective, government funds have been committed to strategic projects addressing grape production and farm gate issues. The Federal government also provides funds through several agencies (e.g. Australian Council for International Agricultural Research) for collaborative research and development projects between organizations in Australia and throughout Asia, designed to assist industry development.

GRAPE PRODUCTION IN CHINA

Grapes (*Vitis vinifera* L.) have been grown in China for more than 2000 years. However, grape production was negligible until the founding of the People's Republic of China.

The area under cultivation of grapes and their production were only about 3,200 hectares and 39,000 tonnes, respectively. The government paid great attention to improvement of the quality of life of the people after the founding of the People's Republic of China.

Following rapid economic growth, fruit production, including grape production increased rapidly in the country, especially since the 1980's. The land area under grapes and their production in 1998 were respectively estimated to be 55.6 and 60.5 times greater than in 1949.

This paper presents the current status of grape cultivation, the development of the wine industry, advances in cultural techniques, the potential for expansion, and some suggestions for the future development of grape production and the wine industry in China.

PRESENT STATUS OF GRAPE CULTIVATION IN CHINA

Following apple, citrus, pear and banana, grape production now ranks fifth in fruit production in the People's Republic of China. The area under vineyards was estimated to be 178,000 hectares in 1998. To examine the evolution of the area under cultivation of grapes after the founding of the People's Republic of China, two rapid development periods can be observed during the eighties and last few years of the last century.

The land area under cultivation of grapes increased steadily from 31,600 to 146,000 hectares during the first rapid development period from 1980 to 1988, with an average increase of about 14,300 hectares per year. Then the grape extents decreased from 1989 to about 24,000 hectares in the following 3 years. Under the influence of the "red wine rush" in the Asian countries, grape growing again became the focus of people's attention in China from the early nineties.

Thus, the area under cultivation of grapes continued to increase again from 1992 to 20,200 hectares by 1998 (about 15 percent more when compared with that of 1997). The area for the other four main fruit crops, however, decreased by 216,800 hectares for apple, 5,540 hectares for pear and 39,000 hectares for citrus.

The annual grape production has increased steadily for the last 20 years. It attained 2,358200 tonnes in 1998, 18.7 times more than in 1979. Most of this production was for table grapes and only about 10 percent for wine making and another 10 percent for raisins. The main areas of expansion are located in the north. The five leading provinces that contributed more than two thirds of the total grape production in the country in 1998 were Xingjiang, Hebei, Shangdong, Liaoning and Henan.

Table. Main Grape-Growing Regions and their Production, 1998

Province	Vineyard Area		Grape Production	
	1000 ha	% of the Country total	1000 tonnes	% of the Country total
Xingjiang	32.0	18.0	555.4	23.6
Hebei	31.5	17.7	404.4	17.1
Shangdong	20.9	11.7	269.0	11.4
Liaoning	14.2	8.0	275.6	11.7
Henan	11.3	6.3	153.0	6.5
Total in China	178.0	100	2358.2	100

Although the Chinese have a tradition of consuming alcoholic beverages, China has been very backward in wine industry development. At the time of the founding of the People's Republic of China, the estimated annual production of wine was about 85 tonnes per year. Although the wine industry has developed rapidly since that time, wine production has contributed only a very small fraction towards the total alcoholic beverage production in the country.

Following a rapid growth in wine production, most of the provinces are now engaged in the industry. More than 400 wine companies and wineries were established in 26 provinces and municipalities operating directly under the Central Government at the end of 1997. The six leading provinces or municipalities which produced more than 10,000 tonnes of wine in 1997 were Shandong (65,500 t), Hebei (23,600 t), Tianjing (17,000 t), Beijing (16,400 t), Anhui (15,700 t) and Henan (14,600 t).

The major areas for growing wine grapes, however, are essentially located in the North of China, such as in the Northeast, including Bohai Bay, Shacheng and Changli (Hebei), Qingxu (Shanxi), Yinchuan (Ningxia), Wuwei (Gansu), Tulufan Basin (Xinjiang), Shihezi (Xinjiang), as well as the ancient course of the Yellow River valley and Yunnan Maitraya.

Among hundreds of grape varieties grown in China, the following are the more important, from an economic point of view:

Table grapes: Niunai (Cow's Nipples), Kyoho, Muscat Hamburg, Thompson Seedless, Longyan (Dragon Eyes) Jingxiu, Zana, Rizamat, Fenghuang No 51, Red globe, Jingzhaojing, Jingyu, Italia and Fujiminori.

Wine Grapes: The main varieties for white wine are Chardonnay, Italian Riesling, Ugni blanc, Chenin blanc, Gewurztraminer, Sauvignon blanc, Semillon, White Riesling and Rkatsiteli. The main varieties for red wine are Cabernet Sauvignon, Cabernet franc, Merlot, French blue, Muscat Hamburg, Pinot noir, Syrah, Carignan and Saperavi.

PRODUCTION OF PLANTING MATERIAL

China is fortunate in not having any phylloxera pest problems. Most of the grape planting material used in the country is therefore propagated by cuttings. In some cold areas, however, vines are grafted on cold-resistant rootstocks, such as Beta (a probable hybrid between *Vitis riparia* and *V. labrusca*) and lines from *Vitis amurensis*.

ESTABLISHMENT OF ORCHARDS

Thorough preparation of the land is essential for successful establishment and good vine growth during the first two to three years because of poor organic matter status in the soil. The land is trenched to a width of 0.6 to 1 m and a depth of 0.8 to 1 m. The distance between the trenches depends on the distance between rows. The soil is mixed with organic matter or other fertilizer and filled back into the trenches before planting.

Young plants are set out in the vineyard in late autumn or early spring before growth begins. The planting season of vines depends upon the climate of the region. To reduce the risk of frost, vines have to be planted in early spring in most areas of north China where it is very cold in winter, while they may be planted in late autumn in the Valley of Yangtze River. The plant spacings used for grapes vary depending on the regions and the training systems. The general information on the spacing and density used for establishing a new vineyard. Planting density varies from about 1,000 to 5,000 vines per hectare.

Table. General Suggestions for Plant Densities to be used for the Establishment of New Vineyards

Region	Trellis Form	Spacing between Rows (m)	Spacing in the Row (m)	Density Vines per ha
North China	Pergola	4.5-5.5	0.75-1.0	1818-2963
	Vertical Trellis	2.5-3.0	0.8-1.5	2222-5000
South China	Pergola	5.0-6.0	1.0-1.5	1111-2000
	Vertical Trellis	3.0	1.0-2.0	1666-3333

CARE AND MANAGEMENT OF VINEYARDS

Pergola and vertical trellis are the more popular forms of training used for grape growing in China. The popular training systems in the north are the multi-arm fan system with no trunk used for vertical trellis and the dragon system with a strong permanent cordon where grapevines usually need to be 'heeled' in winter. In south China or in the regions where grapevines need not be heeled in winter, the vertical trellis is frequently used. The multi-arm fan system with a strong trunk, the bilateral cordon system with two strong horizontal cordons, the standard T bar system and the umbrella system are the most commonly used training systems.

Pruning is one of the important aspects of vine management to obtain consistently good yields and fruit quality each growing season. Dormant pruning is done in late autumn or winter before the growth begins. The one-year-old hardwood cuttings can be left as fruiting canes and are cut back to 5 to 9 buds, and sometimes to more than 10 buds if the "cane pruning" technique is used. For the hardwood cuttings to obtain renewal shoots, they have to be cut back to 2 to 4 buds. Summer pruning is done during early growing season by thinning and pinching.

Several fertilizer applications are usually carried per year in most vineyards. In general, three or four applications of chemical fertilizers are applied after bud break, at flowering, during rapid growth of young fruit, and during the maturation of grape berries.

Nitrogenous and phosphorous fertilizer are usually supplied for the first two or three applications while only potash fertilizer is used during the maturation of berries. Moreover, a high quantity of manure (more than 30 t) is often applied after harvest or in late autumn. The last manure application is very important for obtaining a high grape quality because of the rather low content of organic matter in the soil.

China's continental climate produces hot, usually rainy summers, and very dry and cold winters. It is therefore necessary to supply supplementary irrigation for growing grapes in most regions. Two supplies of irrigation, one before the vines are heeled and the other one after bud break are obligatory to have a normal growth of the vines and to obtain a good yield. Irrigation is not recommended one to two months before harvest, especially for wine grapes. The climatic conditions are very variable in China. For example, it is very hot and humid in summer in the Yangtze valley, but very dry in the region of Xingjiang. Diseases are therefore, found to be different from one region to another.

A few insects can endanger grape production in commercial vineyards. Main grape insects are the comstock mealy bug (*Pseudococcus comstocki* Kuwanna), scale (*Parthenolecanium cormi* Borchs), grape leafhopper (*Erythroneura apicalis* Nawa), grape horn worn (*Ampelophaga rubiginosa* Bremer et Grey), boston ivy tiger-moth (*Seudyra subflava* Moore), grape leaf beetle (*Oides decempunctata* Bilberg), grape erinose mite (*Colomerus vitis* Pagenstecher), grape bunch mite (*Brevipalpus lewisi* McGregor), grape clear wing moth (*Paranthrene regalis* Butler) and grape borer (*Xylotrechus pyrrhoderus* Bates).

HARVESTING AND YIELDS

Only several years ago, grape yields were very high in China, especially in the case of table grapes. Many growers obtained very high yields, as much as 40 to 60 t/ha, and sometimes up to 100 t/ha, but obviously at the expense of grape quality.

Consequently, sugar content in the grapes was too low to be used for wine making and table grapes were non-marketable or sold at very low prices. In recent years, growers paid more attention to grape quality than to yield. The productivity is now severely controlled in most vineyards.

Table. Harvest Date and General Quality of some Wine Grapes Produced in Pindu (Shandong Province)

Cultivar	Harvest Date Day/Month	Sugar (Brix)	Acids (%)	pH
Chardonnay	3/9	20	7.1	3.11
Italian Risling	3/9	18	7.2	3.09
Chenin Blanc	23/9	23	8.0	
Cabernet Franc	2/10	24	8.0	
Cabernet Sauvignon	2/10	23	8.2	
Merlot	2/10	24	7.4	3.20
Syrach	22/9	20	8.5	3.00
Gamay Noir	22/9	21	8.0	
Gewurztraminer	6/9	21	6.0	3.25
Sauvignon Blanc	4/9	21	8.2	3.08
Pinot Noir	22/8	16.5	7.7	3.15

The grapevines usually enter into bearing in the second or third year after planting and mature vines may produce 20 to 30 tonnes per hectare in irrigated vineyards in order to obtain a higher quality of grapes.

MARKETING

Most fruits produced in China are traded in the national marketing system. The volume of exported fruits is very small, averaging about 600,000 tonnes per year or 1.2 percent of total national production. The situation with regard to grape marketing is the same as for other fruits. 650 tonnes of grapes and 1,100 tonnes of raisins were exported in 1997.

This volume of grape exports ranks sixth in fruit exports of the People's Republic of China following apples, citrus, pears, bananas and peaches (nectarines included). However, the volume of grapes imported is much more significant if compared with that exported. Nearly 4800 tones of grapes were imported in 1997, some 7.4 times more than that exported. This import was essentially from USA and most of these grapes imported were the Red globe variety. With regard to raisins imported into China, the volume is less than that exported. Nearly 460 tonnes of raisins were imported in 1997, which amounted to about half of that exported.

Table. Fruit Exports and Imports in China (1997)

	Export			Import		
Fruit Crop	Volume (x1000t)	Value (x1000US$)	Price (US$/kg)	Volume (x1000t)	Value (x1000US$)	Price (US$/kg)
Apples	188.4	77441.7	0.411	11.5	3493.6	0.308
Citrus (oranges, Tangerines etc.)	223.1	7584.00	0.340	13.8	3926.7	0.285
Pears	120.0	52579.3	0.438	0.37	126.1	0.341
Bananas	14.7	5364.3	0.365	549.6	146410.4	0.266
Peaches and Nectarines	3.7	1115.2	0.301	0.016	5.7	0.356
Grapes	0.65	226.0	0.348	4.8	2221.9	0.463
Raisins	1.1	1496.0	1.36	0.46	386.9	0.841
Kiwifruits	0.016	33.0	2.062	0.88	554.3	0.630
Litchis	4.8	6216.3	1.295	1.8	699.9	0.389
Total	601.6	234018.9	0.389	640.4	175694.4	0.274

POTENTIAL FOR GRAPE PRODUCTION DEVELOPMENT

As described above, the establishment of vineyards in China is different from that of other countries in the world and most of the grape production is for table grapes. Although the wine industry has made rapid progress in recent years, the per capita production of wine remains small, as low as about 0.2 L per person. This production is far below the average wine consumption in the world (about 7 L per person) and does not satisfy the current consumption requirements of the Chinese people. This deficiency in production has resulted in the following situation in the country. Firstly, many new wine companies and wineries have sprung up like mushrooms in recent years. There were about 240 wine companies and wineries by 1995, after a development phase spanning over a century in China, but an additional 200 wine companies and wineries were founded only during 1996 and 1997.

The wine production capacity was almost doubled during these two years, mostly in Hebei, Shandong, Yunnan, Gansu, Ningxia, and Guangdong. Secondly, the area under cultivation of grapes increased rapidly all over the country, especially during the last two or three years. This rise in the cultivated area was essentially for wine grapes. Lastly, China experienced a sharp increase in wine imports, especially from European countries, with nearly 5,900 tonnes of wines being imported into China in 1996, which amounted to 7.7 times more than that of 1995. Import of wines continued to increase during the following years, exceeding 39,670 tonnes in 1997 and 49,840 tonnes in 1998.

The change in consumption habits of alcoholic beverages in modern China resulted in the above phenomenon. Chinese have a tradition of consuming alcoholic beverages, especially drinking of spirits distilled from sorghum and maize. This habit has been changing since the 1970's. Beer production in tonnage was more than 50 percent of the national alcoholic beverage production by the end of the 1980's. Since the early 1990's, consumption of wines has become a trend of the times.

More and more people consumed wine instead of spirits and the demand for wine will continue to increase in the following years. China has, therefore, a high potential for grape development, especially for the wine industry. It is absolutely certain that the area under cultivation of grapes will continue to increase in the coming years.

The margin between wine production and demand is still very wide. Nearly nine millions tonnes of wine would be required per year to satisfy the current national requirements if the average consumption of wine reaches that of world per capita consumption.

CONSTRAINTS IN GRAPE PRODUCTION DEVELOPMENT

After continuous rapid development during the past twenty years, China has achieved great success in grape production. However, some problems arose at the same time as this rapid development of grape production in the country. Some of the salient problems have to be resolved if the country is to have a steady development of the grape industry in the future.

Much more attention should therefore be paid to grape and wine quality rather than productivity, although the latter has been much improved during the last 10 years. Long-term planning for grape development in the whole country must be carried out in identifying new varieties, locating growing regions for new vineyards, and developing strategies for increasing production and processing into wine. Production of planting material has to be controlled by the Government for quality control purposes. Following the rapid increase of area under cultivation of grapes, it has been difficult to control the quality of plants by the Government.

Moreover, many plants were imported mainly from European countries, such as 4.5 million cuttings from France in 1998. Monitoring the health of materials has therefore become a new problem for the grape industry in China. Both the grape production and wine industries need to be supported by the Government, research institutions and foreign companies with equipment, investments and advanced technologies.

GOVERNMENT POLICIES AND PLANS FOR RESEARCH AND DEVELOPMENT ON GRAPES

The Government of China has paid considerable attention to grape production and winery development since the founding of the Peoples Republic of China. Two national grape germplasm repositories were set up at Zhengzhou Fruit Research Institute of the China Academy of Agricultural Sciences (located at Zhengzhou, Henan province) and at the Institute of Fruit Research of Shanxi Academy of Agricultural Sciences (located at Taigu, Shanxi province). More than 1,300 varieties were collected in these two national grape germplasm repositories. Some varieties of grapes, mainly for table grapes, have been developed at the same time by Chinese researchers and breeders. Some of the popular Chinese varieties which have been cultivated or are being popularized are listed below:

Table Grapes: Zaomeigui, Zhengzhuo Zaohong, Fenghuang No 51, Jing Zaojing, Shangdong Zaohong, Jingxiu, Jingya, Zizhenxiang, Shengxiu, Jingyu, Fenghou;

Wine grapes: Beichun, Gongliang No 1, Shuangyou, Zuoshan No 1.

Moreover, many significant research findings have been made on identifying suitable agro-climatic zones for growing grapes (especially for wine grapes). Propagation of materials (including micro-propagation techniques), establishment of vineyards, training systems and pruning methods, fertilization and irrigation management, grape quality improvement, and table grape storage are some of the significant advances made in grape research.

The wine production in China is estimated to be over 300,000 tonnes by the year 2000. According to the wine consumption patterns in the last few years, the Government plans to achieve an annual increase of 50,000 tonnes in wine production in the coming 10 years. This means that China would produce 800,000 tonnes by 2010 and about 3,000 to 4,000 ha of wine grapes must be established each year in the future. In order to satisfy this increase of wine production in China, some emphasis would have to be made by the Government on controlling quality of planting material (by establishing "the Propagation

Centre for Grape Materials"), planning a long-term programme for table grape and wine production in the whole country. The establishment of a wine grape industry and increasing the rate of the development of grape production by training farmers and controlling grape quality and wine quality are some of the important aspects of this strategy.

GRAPE PRODUCTION IN THE REPUBLIC OF KOREA

Some wild species of grapes like *Vitis amurensis* and *Vitis coignetiae* found in the Republic of Korea are not commercially important species. Several cultivars grown until now were introduced through China from Europe, and found their way to Asia during the Koryo Dynasty around the 14th century. Modern viticulture began around 1906 when an experimental station was established to test imported cultivars.

The wine industry began in 1910 in Kyongbuk Province. Some cultivars for wine were imported from USA, Europe, and Japan. These have been grown throughout the country since 1960.

The income from grape growing in the Republic of Korea is much higher when compared with that from other fruits. But over-production due to the rapid increase in grape growing areas and the high amount of imported grapes has brought about economic difficulties to local farmers. Presently, grape consumption is increasing because of the high national income and the superior quality of grapes. Grape production will boom as a promising industry if the quality of berries could be improved and the operating cost reduced.

PRESENT STATUS OF GRAPE CULTIVATION

Grape Cultivation Areas and Yields

From 1994 to 1998, the area planted to grapes expanded at an average annual rate of 11.5 percent. Correspondingly, production grew at an average rate of 18.3 percent. The increasing trend of the area and the volume of production can be attributed to many factors, such as increasing consumption as a health food, ease of cultivation compared with other fruit crops.

The area under cultivation of mature vines grew from 1995 to 1999. On the other hand, the area devoted to young vine cultivation decreased somewhat. Generally however, the trend in grape production, yield and consumption is currently increasing.

Kyongbuk province has the largest area planted to grapes (46 percent) followed by Chungbuk (15 percent), Chungnam (11 percent) and Kyonggi (10 percent). Cheju has the smallest area (8 ha) planted to grapes.

Grape Cultivars Growing in the Country

The distribution of grape cultivars is dependent on the prevailing climate in The Republic of Korea. European cultivars cannot be grown without 'heeling' under the soil in winter in the northern areas. Many American cultivars and their hybrids with European cultivars can however, be grown using this method. The 'Campbell Early' cultivar occupies 66.4 percent of the total area cultivated to grapes. The 'Kyoho' variety occupies 14.5 percent, and 'Sheridan' 11.8 percent. These three occupy 93.5 percent of the total area planted to grapes. Various varieties of good quality are currently needed to compete against imported grapes.

PRODUCTION OF PLANTING MATERIAL

Until recently many farmers grew grapevines without using rootstocks. There has been little infestation by grape phylloxera owing to the hot and rainy weather in summer and cold conditions in winter. It is reported that the grape phylloxera was present in some regions in 1912-1913. In 1998, phylloxera emerged in Cheonan and Anseong areas where mostly the 'Kyoho' cultivar is grown, but its spread has been nearly brought under control at present. To protect the grape from phylloxera and increase yield and colouration of berries, many farmers in The Republic of Korea began to use grafted grapevines. In some important regions like Cheonan, Anseong, Naju, Youngdong, and Kimchoen, farmers try to mass produce grafted nursery plants using rootstocks such as 5BB, SO4, 3309, 3306, 5C, and 8B etc. In addition there are some Agricultural Extension Services which plan to produce virus-free plants to supply their farmers.

ESTABLISHMENT OF VINEYARDS

Land Preparation

Precipitation in the Republic of Korea is more than 1,000mm a year, with almost half of it falling during June and July. This makes drainage installation important for vineyards. Considering the resistant level of grape cultivars against winter coldness, farmers are recommended to plant European cultivars in the areas where the minimum temperature is above -15°C and for American cultivars, above -20 to -25°C. Grapevines should be heeled to overcome these critically low temperatures during winter. The cultivars growing in the Republic of Korea have a large range of soil adaptability; hence planting vines on plains and slopes is possible. Generally, farmers prefer to plant grapevines in sandy loam soils. They make agricultural roads between vineyards for power tillers to operate and grow rye cereals or clovers as cover crops to avoid soil erosion.

Planting Season

Grape growers plant vines from November after leaf fall to the end of March when their roots start to revive. Generally, they prefer to plant grapevines in spring to avoid winter injury. Planting nursery stock is recommended before the end of March.

Spacing

Vine spacing is different according to varieties, rootstocks, soil fertility and other management techniques. To establish the Wakeman's system for 'Campbell Early' grapes, it is recommended to have spacings of 3.6m x 2.7m (1,000 vines/ha) on fertile soils and 2.7m x 2.7m (1,370vines/ha) on poorer soils. In the Republic of Korea, every other plant is removed when the distance between rows is less than 1.8 - 2.4m.

In the shelf system for 'Kyoho' grapes, plants usually have a spacing of 1.8m x 3.6m at the beginning and then 3.6m x 3.6m in 4th - 5th year, the spacing of 5 - 6m in the row is acceptable.

CARE AND MANAGEMENT OF VINEYARDS

Training and Pruning of Vines

The shelf system, which has short and long canes according to the diameters of canes, is used for the 'Kyoho' grape variety. The umbrella system is also used, but this form is difficult to control when vines are fully grown. This should be changed into a simple and an effective form, which is appropriate for plastic culture.

It is advisable to get 1cm-thick bearing mother canes for 'Kyoho' grapes. Each shoot of 'Campbell Early' grapes has an average of 1.5 clusters. Two buds are needed from each bearing spur branch and a shoot is chosen out of the lower part from the two buds.

Application of Manure and Fertilizers

Farmers apply fertilizers separately. Basal fertilizers are applied after leaf fall in November up to pre-sprouting stage in March. Grape growers apply slow releasing manure and a little more fertilizers to provide nutrient element requirements. They supply all of the phosphorous and lime at the same time. More fertilizers are sometimes applied from the end of May to the beginning of June. In this case, 20-30 percent of nitrogen and 40 percent of potassium should be applied at the berry enlargement stage. Sometimes farmers supply a small amount of fertilizers after harvest for recovery of vine vigor. Urea application on leaves to supplement nitrogen is also useful. After manure and fertilizers have been evenly applied on the surface of the soil, a tiller is used to plow and turn the soil.

Weeding

Too much rain during the vine growing season makes grasses grow well. Many farmers use herbicides 3-4 times a year to control weeds. Some farmers are also increasingly using cover crops like rye cereals and clovers to control soil erosion and weeds.

Supplementary Irrigation

The Republic of Korea has sufficient rainfall (1,000~1,300mm per year), but 40-50 percent of this precipitation occurs during July and August. Sometimes, irrigation is required from the beginning of May to the end of June. Precipitation from September to October is not enough; hence supplementary irrigation is necessary.

Pests and their Management

The common pests in the Republic of Korea are grape tiger longhorn, grape clearwing moth, grape leafhopper and yellow tea thrips. In 1998, grape phylloxera infected areas were found in some regions, but the infected areas are no longer increasing at present. Generally, farmers use insecticides about 10 times a year.

However, some farmers apply these chemicals less than 3-4 times a year by using plastic houses.

Diseases and their Management

The common diseases in the Republic of Korea are bitter rot, leaf spot, downy mildew, powdery mildew, gray mold and anthracnose. In 1998 and 1999, bitter rot, downy mildew and anthracnose were a little severe due to heavy rains lasting until harvesting time.

Physiological Disorders

Poor sprouting and bud necrosis at the early growth stage, berry shattering, and cluster stem dehydration are conditions observed in relation to cluster formation. Sometimes, excess and deficiency of nitrogen, potassium, magnesium and boron are common. Many disorders occur under abnormal weather conditions.

There is sunburn on berries and leaves, berry cracking and water berries. Very severe berry cracking and water berries occurred in 1998 and 1999. The search for solutions to these disorders is being undertaken through several research studies.

Other Factors such as Grape Quality Improvement

Plastic culture is the best method to improve berry quality, and so it has been popularized.

Many growth regulators like GA, KT-30, and ethephon are being used to induce seedless grapes, berry enlargement, hastening and delaying harvest, cluster elongation, and improvement of berry setting. Sometimes improper application creates problems. Many farmers use products made commercially, and also produce their own formulations themselves even if their efficacies are not proven.

HARVESTING AND YIELDS

Grape yield is increasing annually due to the increase of grape cultivation area and advanced grape growing skills. Grape production in 1994 was 211,930 tonnes and reached 397,784 tonnes in 1998. In 1999, the production of grapes was estimated at around 470,000 tonnes because of increased growing area of mature grapevines.

Growing grapes is much easier than other fruits; therefore farmers can easily get a good income. Berry shattering occurs every year especially in 'Kyoho' grapes, but this does not seriously affect grape production. In spite of bad weather conditions, farmers keep having good income because of excellent berry quality under plastic houses and their growing skills. The consumption of grapes has also been increasing steadily.

MARKETING

Grapes are sold through Agricultural Cooperative Associations, Producers' Cooperatives, by the roadside, and at department stores. The Republic of Korea exports 'Delaware' grapes to the South East Asian markets and sometimes to Japan, but also imports wines, juice and table grapes. Table grapes like 'Red globe', 'Thompson seedless', and 'Flame seedless' are imported from Chile and USA, and since the change of the import policy in 1996, imports are increasing. It is assumed that much more table grapes will be imported after the conclusion of the Free Trade Agreement in Agriculture between the Republic of Korea and Chile. At the same time the local price of grapes has declined since 1994 due to over-production.

POTENTIAL FOR GRAPE PRODUCTION DEVELOPMENT

The Government of the Republic of Korea recommends that farmers do not plant grapevines anymore. But some farmers continue to grow vines because they do not have other alternative crops.

Farmers try to produce good quality grapes and they are making grape juice and wines for themselves. In some regions, farmers grow a hybrid cultivar between a wild Asian variety and an American cultivar to produce a popular wine and juice. But its quality is variable according to the different methods used by farmers to produce wine.

CONSTRAINTS IN GRAPE PRODUCTION DEVELOPMENT

Growers are supposed to apply chemicals about 10 times a year, but this frequency is reduced to less than 3 - 4 times by using plastic houses. Recently, the grape crop has had several physiological disorders like berry shattering, berry cracking and water berries under abnormal weather conditions. It is necessary to process surplus grapes and export more fresh grapes. This may lighten the farmers' burden caused by excessive production. Grape growing farmers should produce grapes of good quality under plastic houses and install irrigation systems. They also need some technical help like soil testing and continuous education of growers on new techniques.

GOVERNMENT POLICIES AND PLANS FOR RESEARCH AND DEVELOPMENT ON GRAPE PRODUCTION

The Government discourages farmers from growing grapes and no financial help is provided. Despite this situation, many researchers are still conducting studies on the pests, diseases and physiological disorders of grapes. Some courses are organized by Government organizations to train farmers on these aspects.

PRESENT STATUS OF GRAPE CULTIVATION

In Myanmar, grapes are grown mainly for the fresh fruit market. Grapes are grown not only in state farms but also in private farms. The cultivated area of grapes is shown in Tables 1 and 2.

Table. Grape Area in Private Farms

State & Div. Township	Cultivated	Variety (ha)					
	Area (ha)	Bokhari	Muscat	Pizzutello	Kuato	Cardinal	Alexandria
Mandalay							
Meikhtila	107.65	85.80	11.33	10.52	-	-	-
Yamethin	373.13	267.10	57.47	48.56	-	-	-
Pyawbwe	206.80	153.79	29.94	23.07	-	-	-
Tatkone	0.40	-	-	0.40	-	-	-
Shan							
Tachileik	40.49	-	-	-	40.49	-	-
Taunggyi	15.61	13.36	-	-	-	1.25	1.00
Magway							
Pakokku	1.21	1.21	-	-	-	-	-
Rakhine	2.02	2.02	-	-	-	-	-
Total	**747.31**	**523.28**	**98.74**	**82.55**	**40.49**	**1.25**	**1.00**

GRAPE PRODUCTION IN MYANMAR

Grapes were planted in Myanmar on an experimental scale in 1959 at Sebauk State Farm, Kyaukpadaung Township, and Myingyan district, in the dry zone of Central Myanmar. As it was shown that cultivation of this crop was feasible in this climatic zone, grape was extended to Kyaukpadaung, Meikhtila, Nyaung Oo, Yamethin, Pyawbwe and Pakokku townships.

These townships are very dry areas of Central Myanmar. The climatic conditions of grape growing areas of Myanmar are probably different from that of other grape growing countries.

Therefore, cultivation practices for grapes in Myanmar may not be the same as in other countries. Furthermore, there was only scanty research information available concerning grape cultivation in the 1950's. Consequently, the practices of cultivation were developed through experience by technicians of Sebauk grape farm.

Grapes naturally flower in April-May at the beginning of the rainy season and consequently there is flower drop and less fruit due to the rain, also the fruits that set in the rainy season have lower sweetness. Thus pruning systems were developed to make fruits set in the dry period, which is between February and March.

PRODUCTION OF PLANTING MATERIAL

In Myanmar, use of cuttings is the only method for propagation of grapes. However, budding and grafting techniques were tested very recently in state farms but promising results have not been observed yet and experimentation of these techniques is still continuing.

Propagation of Grapes by Cuttings

Six to twelve month old hardwood cuttings are produced from established vines. The cuttings are taken in mid-October from healthy mature vines. The pencil-width mature canes 20-30 cm long are cut with three or four buds in each. The cane is cut at 2.5 cm away from the nodes, and the lower cut is slanting for easy planting.

The cuttings are taken from the canes already pruned and removed from the vines. The cuttings are dipped in sulfenatte suspension (1:300) for about 5 minutes and put in boxes. In the box, the cuttings are covered with a soaked gunny sack to maintain moisture and the boxes are then transported to the nursery sites.

Propagation at Nursery Site

The beds are prepared at least one month ahead of the cuttings being taken. Beds of 1 meter wide, 30 cm deep and 7 meters long are demarcated. The path between the beds is about 45 cm. The soil of the beds is loosened and left for 5 days for sun drying. About 2 baskets of farmyard manure (40 kg) is mixed with the soil in each nursery bed. Then the beds are watered for a few days. If the soil is hard and of high clay content, sand is added to loosen the soil for better aeration.

On the bed, planting points of 7.5 cm x 7.5 cm apart are prepared and slanted planting holes 15 cm deep are made by using a stick with a diameter of about 5 cm.

The cuttings are planted in the holes and the soil around each cutting is pressed down by foot. The nursery beds are watered once a day in the evening and hand weeding is done as necessary. If the weather is too hot and sunny, partial shade is given for the nursery beds. One month after planting the cuttings in the bed, the new shoots come out and they are transplanted into polybags.

The soil to fill up the polybags is prepared by mixing topsoil, sand and organic matter in a ratio of 10:5:1 ratio by volume. The prepared soil is filled in ploy-bags of 18 x 30 cm size and the soil is filled up to half of the bag.

The cuttings are dug up taking care not to damage the roots, and the roots are then washed out. The cutting is planted in the bag and soil is filled and pressed down gently. Polybag cuttings are put under shade and watered twice a day and placed under close supervision for controlling pests and diseases as necessary.

ESTABLISHMENT OF VINEYARDS

Land Preparation

Normal land clearing is done by under brushing, cutting, burning, stacking, re-burning and stumping. When the land is cleared, ploughing with tractors is done, and then harrowing and leveling of the field is carried out.

When land clearing is finished, planting points are marked by pegging. The spacing depends on the variety and training system adopted. There are three training systems, Arbor, Kniffin and Y- type or Cordon.

Spacing

Spacing depends upon soil fertility, variety and training system. Normal spacing used in Myanmar is shown according to the training system.

Hole Making

Holes of 0.3 m x 0.3 m x 0.3 m are dug and the topsoil and subsoil is separated. Each hole is filled with 40 kg farmyard manure, 30 kg compost and topsoil. The manure is augmented with 4 kg bone meal and 2 kg of wood ash. Holes are irrigated two times daily and planting is done 15 days after irrigation. Posts are constructed according to the training systems used.

Planting

Grapes are planted from mid-October to February, which is the best period for establishing grapes in Myanmar. Normally planting is done in the evening. A small hole 15 cm wide and 30 cm deep is dug at the centre of the planting hole. The plastic bag is removed without damaging the soil core of the seedling. Then it is planted in the small hole gently and the soil is pressed down well. The level of the soil core of the cutting must be the same as that of the planting hole and deep planting or shallow planting should be avoided.

CARE AND MANAGEMENT OF VINEYARDS

The plants are irrigated at least once a week depending on the moisture of the soil.

- Cyan or Furadan is used to prevent termite damage.
- Hand weeding around the plants is done as necessary, weeding in inter rows is also done.
- Inter-cultivation is carried out occasionally using harrows or other cultivators.
- New shoots usually come out about one month after planting.
- When the vine is 30 cm long, it is tied to the poles already erected near the plant.
- The strongest shoot is left and other shoots are pruned and any new shoots are also pruned.
- At three months after planting, the weak plants that do not show normal growth are removed and replaced.
- Gap filling for dead plants is done if necessary. To get the required plant population it is important to plant in time with similar sized plants. If planting is late, those plants will not be able to compete with surrounding plants.

TRAINING AND PRUNING OF PLANTS

Training

There are three kinds of training systems in Myanmar. They are Arbor, Kniffin and Y-type or Cordon. Training systems depend on the growth habit of the variety and climatic conditions of the place where grapes are going to be planted. If the grape variety has good growth and branching habit, Arbor system is used. When grape is grown in areas of less rainfall the Arbor system is preferred. As the grape growing areas of Myanmar are in the arid region of Central Myanmar, most of the farmers use the Arbor system. Kniffin and Y-type systems are used only in the State farms as an experimental study. If the area has higher rainfall and humidity, the Kniffin and Y-types should be used.

Arbor System

The height of the Arbor is 1.8 - 2.4 m, however 1.8 m is preferred because of convenience in pruning and other cultivation practices. The buds on the trunk of 1.8 m height are rubbed off leaving only the main stem. When the main stem reaches the top of the Arbor, it is cut out at the tip of the trunk. Then the shoots come out from the tip of the trunk, but they are not pruned unless they have abnormal growth. When the canes have grown long enough, topping is done at 1.5 m from central trunk.

Kniffin

In Myanmar, the four-cane Kniffin system is being tested. The trunk is cut at a height of 0.75 m and three canes are left to grow. Two canes are allowed to grow on the first wire, and one cane to grow up until the second wire where it is topped again to let the new canes come out. The canes are tied to the wires. The lateral four canes are cut at 0.9 m from the trunk. The main trunk is supported by a pole until the second year of growth.

Y-type or Cordon

The trunk is cut at a height of 1.2 m and the strongest two shoots are allowed to grow. These two canes are tied to the bottom wire of the Y, and are cut 1.2 m from the trunk. The new shoots that come out of the canes are allowed to grow, and are tied to the upper wires of the Y.

Pruning

There are two kinds of pruning practiced in the country - light pruning and heavy pruning. Light pruning is done occasionally to prune unwanted canes and leaves for better light penetration and removal of diseased canes. Heavy pruning is done two times a year. The first pruning is done in mid-October to mid-November and the second pruning in mid-March to mid-April.

In the first year of planting, some flower clusters that come out after the first pruning (mid-October - mid-November) are removed, and only a few clusters are left for evaluation of the fruit quality. Otherwise heavy fruiting in the first year will weaken the growth of the young vines. Commercial harvesting is done from the second year onwards.

If the trunk is wounded by some diseases or other means, it is cut leaving the healthy portion of the trunk and new shoots are allowed to grow and are trained again.

HARVESTING AND YIELD

Irrigation is stopped one month before harvest to reduce water content of the berries, which will increase sugar percentage of the fruit. Manual harvesting is done when the berries are ripe enough. The peak season of production is February to May.The total production of grapes in Myanmar is 8,310.09 tonnes from an area of 747.31 hectares giving an average yield of 11.12 t/ha. The low yields are due to lack of knowledge of the growers about the training and pruning practices and disease control measures. The poor quality of grapes in Myanmar is due to diseases, especially powdery mildew and variety.

MARKETING

There is no export or import of grapes in Myanmar. The local demand is always higher than production. The harvested grapes are sent to the two main purchase depots in Mandalay (Upper Myanmar) and Yangon (Lower Myanmar). From those main centres, grapes are sold at retail prices in small shops and roadside stalls.

POTENTIAL FOR GRAPE PRODUCTION DEVELOPMENT

There is potential for the development of grape production by area extension and yield improvement. Grapes can be extended in the divisions of Central Myanmar such as Mandalay, Magway, and Sagaing Divisions.

Moreover, there would be a substantial increase in production of grapes if the growers are educated in the training and pruning techniques and given high yielding quality grape varieties.

CONSTRAINTS IN GRAPE PRODUCTION DEVELOPMENT

There are some constraints for yield increase and area extension of grapes in Myanmar. Lack of technical knowledge such as training and pruning is a major factor that contributes to low yield. Fungus diseases are also difficult to control because the outbreaks of these are dependent upon weather conditions. Therefore, variation of yield losses due to diseases is observed every year. Lack of high yielding varieties of good quality is also a factor contributing to low yield.

The most important factor for extension of grapes is the high investment cost for the establishment of vineyards. The establishment costs for the first year are shown below.

Training System	*Cost per hectare*
Arbor	3,200,000 kyats
Kniffin	370,000 kyats
Y-type	490,000 kyats

GOVERNMENT POLICIES AND PLANS FOR RESEARCH AND DEVELOPMENT OF GRAPES

It is realized that research investigations are essential for the development of the grape crop. Therefore, the Government is planning to promote research programmes. At present, some simple research programmes have been started in the State Farms.

- Germplasm collections.
- Comparison of training systems.
- Live-post substitution in training systems.
- Compost and fertilizer trials.
- Mulching techniques.
- Irrigation systems.
- Methods to increase fruit quality.
- Off-season fruit production.
- Top-working techniques.

- Stock and scion relationship.
- Effects of GA.3.
- Weed control techniques.

GRAPE PRODUCTION IN VIET NAM

In Viet Nam, grapevines can be seen everywhere. For a long time, the people in Hanoi Capital have grown grapevines around their houses for ornamental purposes and for shade; but these grapevine trellises have produced small berries and bunches with very sour taste and low food value. Grapes are now grown commercially only in the South. Their quality is not really ideal but is not inferior to other fruits (Vu Cong Hau, 1997).

In other parts of the world viticulture has existed for thousands of years, but in Viet Nam it has just begun to be developed in recent years. Since its economic value is superior to that of other crop plants, areas under grapes in Viet Nam have quickly increased since the early 1990's when only table grapes were the main article of commerce from grape cultivation.

The research work on grapes in Viet Nam has mainly been carried out by the Cotton Research Centre, which has a Food and Fruit Crops Department serving local agriculture. To understand more about the viticulture situation in Viet Nam and help the vine growers to solve production problems in varieties and techniques, since 1989 the Cotton Research Centre (an agricultural scientific research organization of the Government based in Ninh Thuan province) has carried out research on some urgent problems for the development of viticulture.

PRESENT STATUS OF GRAPE CULTIVATION

Major Vine Growing Regions

Growing grapes for commercial purposes began in the late 1970's in the North of former Thuan Hai province (including Ninh Thuan province and Tuy Phong District of Binh Thuan province).

In Northern provinces such as Bac Giang, Ha Tay, Hai Duong, Quang Ninh and Vinh Phuc, vines have begun to be planted on a small scale for experimentation. Hence, the research work has only been carried out in the vine growing region of Ninh Thuan province.

Area Under Grapes

The statistical data on the areas under grapevines in Viet Nam are presented in Table below.

The data show that grapes have mainly been grown in Ninh Thuan and Binh Thuan provinces. Today, the areas under grapevines are mainly located in Ninh Thuan province (occupying 90 percent of the total area) and in Binh

Thuan province (occupying 9 percent of the total area under grapes in the whole country).

Table. Areas under Grapevines in Viet Nam in Recent Years

Provinces	Areas under Grapevines (hectares)				Notes
	1985	**1990**	**1995**	**1999**	
A. In the North:					Both table grapes and grapes for processing (wine and raisins)
1. Bac Giang	0	0	23	3	
2. Ha Tay	0	0	16	10	
3. Hai Duong	0	0	17	0	
4. Quang Ninh	0	0	7	0	
5. Vinh Phuc	0	0	12	12	
B. In the South:					
1. Binh Thuan	0	0	100	250	Only table grapes
2. Ninh Thuan	26	306	1952	2400	
Total (hectares)	**26**	**306**	**2127**	**2675**	

Up to now, statistical figures on the production of grapes in Viet Nam have not been given by the State Statistical Department because the area under grapes is very small in comparison to other crops.

Variety Information

In the middle of 1980's, the four grape varieties that were popular in former Thuan Hai province were Muscat blanc, Alden, Ribbier and Cardinal (Vu Xuan Long et al, 1992). In the 1990's, only the Cardinal variety became popular. This variety has some advantages such as high yield potential, difficult separation from pedicels, and short growing duration in comparison with Muscat blanc, Alden and Ribbier varieties. Duration of each cropping cycle is about 85-95 days from pruning to harvesting, with one month resting of vines before pruning again. Such a cycle needs about four months duration and three harvests per year can be possible.

Viet Nam did not introduce any new grape varieties for nearly 30 years. Hence, both producers and consumers have not been able to select other varieties at will and they have had to depend on the Cardinal variety to meet their urgent needs for local consumption.

Climatic Conditions in Grape Growing Regions

The rainfall in the Northern grape growing regions is much higher than that of Ninh Thuan. Annual rainfall in these provinces varies from 1,500 to 1,900 mm (Appendix 1). High rainfall is the main impediment that caused areas under grapes in the North not to increase but to be reduced, which was also based on experimental studies carried out for some years. Ninh Thuan is a dry region with a mean temperature of 24-28° C, a low relative humidity of 72-85 percent, high sunshine hours per month ranging between 130-287 hours, an annual average rainfall of 750-900 mm, and a lesser number of rainy days per year (Appendix 2). These climatic conditions are fairly favourable for the growth and development of grapes.

Soil Conditions in Ninh Thuan Grape Region

The results from a survey on the nutrition status in the soil of 800 hectares in Ninh Thuan showed that the soil conditions were very suitable for the growth and development of grapes, which included neutral to low acidity, high availability of phosphorous and potassium, and medium level humus content.

PRODUCTION OF PLANTING MATERIAL

Many years ago, there were no Government or private nurseries producing planting material of grapes in Viet Nam. Hence, farmers had to produce planting material of grapes for themselves by using cuttings and a few by layering.

Today, there are three organizations that have been tasked to produce planting material of grapes for farmers, namely the Cotton Research Centre, Ninh Thuan Domestic Animal and Crop Breeding Centre and Binh Thuan Socio-Economic Development Centre. All of them are researching on rootstocks to identify suitable planting materials for different soil and climatic conditions of the various grape growing regions. Besides, the Cotton Research Centre is conducting studies to propagate grapes by the in vitro method.

ESTABLISHMENT OF VINEYARDS

Land Preparation

Some operations that need to be carried out before planting include deep plowing and harrowing to improve good porosity and aeration of the soil. The land is then divided into plots following the general gradient of the land to facilitate appropriate irrigation and drainage layout. This is followed by fine leveling of the surface of each plot, digging holes, and filling with a mixture of topsoil, phosphate fertilizer and lime (if necessary) before commencement of planting.

Planting Season

In Ninh Thuan and Binh Thuan provinces, the farmers usually plant grapes at the end of the rainy season and at the beginning of the dry season. The most suitable planting time is in December and January.

Spacing

Grapes are planted at a spacing of 2.5 x 2.0 m, which gives a density of about 2,000 plants per hectare. In Ninh Thuan and Binh Thuan grape growing regions, the farmers have only used the overhead arbor system for training vines. The height of the trellis depends on the convenience of the worker.

CARE AND MANAGEMENT OF VINEYARDS

Training and Pruning of Plants

Training

In Viet Nam, vine branches are trained into a cobweb-like design. When vines climb to the top of the trellis (about 2-3 months after planting), the tops of main stems are cut in order to promote 2-3 main arms or branches. After 1-1.5 months, the tops of these arms are pruned back to induce 2-3 secondary branches, and canes are spaced at convenient distances along the secondary branches to achieve the required network. Ninh Thuan farmers have a habit of pruning bearing canes for the first harvest after allowing vines to grow for about 8-10 months.

It is necessary to tie branches to the trellis and pluck unwanted axillary buds in time before the blooming period. Vine branches are usually tied along the horizontal plane on the surface of the trellis for better flowering. Plucking axillary buds is regularly done in order to create good conditions for growth and development of the main branches.

Pruning

Vines in Ninh Thuan can be induced to produce at any time of the year (2.5-3 harvests a year), so grape pruning time is possible all the year round. A grape season lasts about 4 months, and 25-30 days after each harvest vines are pruned again for the next crop.

The 8-month old branches are usually pruned in winter-spring for the first crop and every 4 months thereafter. Pruning positions on the canes are from the eighth to the twelfth nodes for the strong branches, and from the fourth to the sixth nodes for the weak canes.

Application of fertilizers

The results from a survey on fertilizer use are presented in Table below.

Table. Fertilizer Doses for Grapes in Ninh Thuan (Cotton Research Centre, 1992)

Kinds of Fertilizers	Total (kg/ hectare/crop)	Percentages of Weights on Different Stages (%)		
		Before Pruning	Pruning-Flowering	Flowering-Ripening
1. Humus	25.300 ± 7.850	100	-	-
2. N	539 ± 68	54.76	26.19	19.05
3. P_2O_5	339 ± 56	65.30	10.20	24.68
4. K_2O	288 ± 73	45.25	11.21	43.53

Application of nitrogenous fertilizer was mainly carried out before pruning (55 percent N), from pruning to flowering stage (26 percent), and 19 percent at berry enlarging stage.

Almost all phosphorous fertilizer was given before pruning, the rest was applied at the bud, flower and berry enlarging stages in the form of mixed fertilizers. However, it was observed that fertilizing nearly 25 percent of the total amount of phosphorous fertilizer in the berry growing stage was rather high.

Potassium fertilizer used for grapes before pruning was 45 percent and 44 percent for the berry enlargement and colour break stages.

Weeding

Average vineyard size of each household is very small, usually about 0.1-0.4 hectares. Hence, controlling weeds is only carried out by hand. In recent years, roundup herbicide has begun to be used. At present, the Cotton Research Centre has been carrying out some research on polyethylene film mulching as a measure for control of weeds in grapes.

The results from the survey on farming techniques showed that 70 percent of vine growing households turned up the soil at least once per crop and others did it once per year. Turning up the soil is often carried out after the last harvest together with the incorporation of organic and inorganic fertilizers before the next pruning.

Supplementary Irrigation

According to Ninh Thuan farmers, irrigation and fertilization are the main inputs for ensuring high grape productivity and quality. Irrigation is only necessary in the dry season but not important in the wet season. Farmers have often used the flooding method of irrigation on the surface of grape beds. The quantity of water and the irrigation frequency together with inorganic fertilizer application are the most important practices of all the field operations. Vineyards are usually irrigated according to a 10-15 day schedule, but in sandy soils it is shorter, usually about 5-7 day intervals.

Insects and Diseases

Insects

Vines are attacked by some destructive insects such as mealy bug *(Ferrisiana virgata)*, thrips (Thrips sp.), red spider mites (*Eotraniclus carpini*), and yellow spider mites etc. These pests usually exist on vines and cause great damage in dry seasons.

The main kind of chewing pest is the army worm (*Spodoptera exigua*). It bites young leaves, buds and flowers. It occurs and causes serious damage in dry seasons.

Diseases

Grapes are infected by several diseases that are economically important.

Downy mildew disease caused by *Plasmopara viticola* fungus infects mostly young and fully grown leaves. The population dynamics and damage caused by this pest indicate that the disease is persistent in vineyards and causes damage all the year round, especially in the rainy season (May to November).

Bunch rot disease caused by *Diplodia natalensis* fungus infects the peduncles of bunches at flowering and young berry stages. Hence, it is very difficult to control this disease when vines are infected which causes serious losses to grape yields. Population dynamics and nature of damage indicate that this disease seems to occur nearly all the year round, especially in the rainy season (May to November).

Powdery mildew disease caused by *Uncinula necator* fungus infects berries and old branches. Population dynamics and damage show that this disease exists in vineyards and develops and causes damage all the year round, especially in dry seasons (December to April).

Rust disease is caused by *Kuehneola vitis* fungus, which infects fully grown leaves. This disease often develops and debilitates vines only in the rainy seasons (June to November).

Table. Some kinds of Pests and Diseases Recorded on Grapevines in Viet Nam

Common Name	Scientific Name	Infected Part
Diseases:		
Downy mildew	*Plasmopara viticola*	Young and full leaves
Bunch rot	*Diplodia natalensis*	Peduncles of bunches
Powdery mildew	*Uncinula necator*	Berries and old branches
Rust	*Kuehneola vitis*	Full and old leaves
Pests:		
Army worm	*Spodoptera exigua*	Leaves, squares and flowers
Mealy bug	*Ferrisiana virgata*	Branches
Thrips	*Thrips sp.*	Leaves, Berries, Branches
Spider mites:		
Red spider mites	*Eotraniclus carpini*	Leaves
Yellow spider mites	*Unknown*	Leaves
Nematodes:	*Unknown*	*Roots*

Pest and Disease Management

As mentioned above, the Cardinal variety in Ninh Thuan has degenerated and possesses weak pest resistance. Moreover, haphazard pruning at any time of the year without any organized cropping system always causes conditions favourable for pests and diseases to develop and cause damage. Hence, controlling them becomes one of the important factors that affect the yield and economic effectiveness in viticulture programmes.

In the past, lime-sulphur, copper sulphate and bordeaux mixture were mainly used to control diseases of grapes. These chemicals were cheap but had to be sprayed many times per crop. When new foreign-made chemical fungicides and pesticides were used to control diseases and insects of grapes, some remarkable results were obtained including fewer spraying times, long-lasting effect from each spraying, and lower labor costs. Since 1989, the Cotton Research Centre has screened many fungicides and recommended some of them to farmers, including Rovral, Rozin, Ridomil MZ, Ridomil combi, Score, Topas, Anvil, Antracol, Copperhydroxide, Tilt, Baycor, Curzate, and Bayfidan, etc.

For army worm (*Spodoptera exigua*), it has been found that it is very difficult to control this kind of pest with chemical insecticides. However, researchers found it easy to control army worm with a biological product, Nuclear Polyhedrosis Virus (NPV_{Se}) made at the Cotton Research Centre.

Nematodes, such as the vine-damaging nematode, were mentioned by scientists of the Cotton Research Centre some years ago. However, this problem has not yet been studied due to lack of budget resources.

HARVESTING AND YIELDS

Table. Grape Yield (t/ha) Based on Different Age of Vines and Crops (By Vu Xuan Long, et al. 1992)

Plant age (year)	1992 Spring-Summer Crop (Pruning in April and in May)	1992 Autumn-Winter Crop (Pruning in Sep. and in Oct.)	1992-1993 Winter-Spring Crop (Pruning in Dec. and in Jan.)	Total of 3 crops/year
1	-	-	8.00	8.00
2	10.55 (4.75	2.67 (3.46	18.67 (4.71	31.89 (10.90
3	20.00 (11.32	8.10 (7.19	20.61 (5.87	40.71 (22.20
4	15.56 (4.45	8.70 (5.85	16.73 (3.29	41.33 (9.82
5	14.90 (4.70	10.24 (6.61	16.30 (2.98	41.44 (13.06
6	9.38 (3.74	6.13 (2.93	10.62 (4.08	26.12 (5.35
7	22.50	5.00	5.75	33.25
12	5.33	7.33	-	12.66
Average grape yield (metric tonnes/hectare/year)				29.43

Average grape yield (metric tonnes/hectare/year) 29.43

The following observations have been made:

- Grapes in Ninh Thuan age fast and become senescent due to severe exploitation of vines. Normally, the vine growers have to get rid of old vines for replanting after they are cropped for about 10 years. This seems to be a bad way of forcing culture (pruning and harvesting 2.5-3 times per year) and the vines consequently degenerate fast.
- The grape yield of the Autumn-Winter season is the lowest in a year, which is only about 50 percent in comparison with other seasons.

The main reason for low productivity is that, in the pruning period of this season (in September and October), heavy and continuous rains often appear, which cause root rot and disadvantages for the blossoming process.

The results from a random investigation of 50 vineyards of different plant ages in 1995 and 1996 showed that mean grape productivity in Ninh Thuan was over 30 tonnes per hectare/year. Planting grapes could bring a profit of over VND 100 million per hectare/year. Grape yield is high and stable in Winter to Spring or Spring to Summer seasons but low and unstable in Autumn to Winter (Appendix 3).

MARKETING

Up to now, most grapes produced in Ninh Thuan and Binh Thuan provinces have been consumed in the form of table grapes, and a small quantity of grapes with very poor quality are processed into local juice and wine. Grapes are transported everywhere in Viet Nam by dealers and middlemen.

At present, Viet Nam is importing 2 kinds of table grapes from abroad. Consumers are aware that they are produced in the United States, Australia or Thailand.

CONSTRAINTS IN GRAPE PRODUCTION DEVELOPMENT

In 1998 and 1999, like other provinces in the centre of Viet Nam, Ninh Thuan was affected by natural disasters (storms and floods) that caused serious damage to viticulture.

The total area under destroyed vineyards by storms and floods was 1,487 hectares (occupying more than 50 percent of areas under grapes at the beginning of 1998). Storms and floods weakened the crop that survived and it may take a long time for them to recover.

Due to the economic advantages of viticulture, some farmers have diversified from other crops to grow grapes, even in double-cropped rice fields or in the sandy soils near the coast. Hence, there were some vineyards established in unsuitable soils such as in sunken terrain which floods frequently, soils with high acid content, as well as in very infertile soils. In order to get reasonable harvests from such soils, farmers had to invest more in comparison to other areas.

Heavy rains often occurred at the end of the year (August to November) together with high temperature, when some diseases develop very quickly. Large quantities of fungicides have to be used in this period.

Up to now, cultivation techniques for grapes have not been properly studied. Farmers have had to toil in their vineyards in order to gain experience for themselves. As a result, the harvests are unstable, uses of fertilizers are excessive and consequently, production tends to suffer.

GOVERNMENT POLICIES FOR VITICULTURE DEVELOPMENT

The Government of Viet Nam has issued several general policy directives for development of fruit crops all over the country, including grapes. These policies include encouraging farmers to change crop planting composition to diversify agriculture and farm products, providing loans to farmers at low interest rates, and constructing new water reservoirs as well as main channel systems for agricultural production. In recent years, Viet Nam Agriculture and Rural Development Ministry has made initial investments for research and agriculture extension including surveys on grape cultivation techniques for farmers, forming agro-extension models for planting new varieties, as well as IPM programmes. These investments are, however, too small in comparison with the grape growers' needs, especially on production techniques.

Since Ninh Thuan province was re-established in 1992, the provincial authorities have affirmed that the development of viticulture is one of the key tasks in its socio-economic programme. Hence, they have begun to show their concern for the development of viticulture by investing more for research on variety collections, surveying the suitable areas for grapes, and building wineries for processing, etc. However, these investments are still insufficient in comparison with production needs.

Research Work on Grapes in Viet Nam

Research on Vine Collections

There are four organizations in Viet Nam maintaining and researching on vine collections. These are the Cotton Research Centre, Hanoi Vegetable and Fruit Research Institute, Ninh Thuan Domestic Animal and Crops Breeding Centre and Binh Thuan Social and Economic Development Centre.

At the Cotton Research Centre these varietal studies carried out from 1994 to 2000 have included the collection and introduction of 61 accessions. Among them there are 12 domestically collected accessions and 49 accessions introduced from abroad (France, United States, Australia, India, Thailand, China and Germany) including 34 table grape accessions, 25 wine grape accessions and 3 rootstock varieties. Through the results of research on this collection, some specific promising accessions were identified and presented for further research. Table grape accessions selected are NH-01-08, NH-01-48 and NH-01-60; wine grape accessions selected are NH-02-04, NH-02-10 and NH-02-17 and the raisin grape accession is NH-01-11.

At Hanoi Vegetable and Fruit Research Institute, 5 accession wine grape collections introduced from France have been studied since 1995. As a result, Vilard noir accession has been identified as a variety having the highest productivity and the best quality (By Nguyen Quoc Hung, Tran The Tuc, Vu Manh Hai and Stephen Desmazieres, 1999). Besides, this Institute has just introduced some additional vine accessions. There are 28 accessions in the collection here at present, including 16 table grape and 12 wine grape accessions.

At Ninh Thuan Domestic Animal and Crop Breeding Centre, the grape collection mainly contained wine grape accessions introduced from Australia from 1995. In 1996, the Germans supplemented some table grape and rootstock accessions to this collection. Now, there are 33 accessions in the collection including 17 table grape accessions, 13 wine grape accessions and 3 rootstocks. From the result of the research on the vine collection, Black Queen table grape variety has been released for screening on an experimental and pilot production scale.

At Binh Thuan Social and Economic Development Centre (established by the Germans together with local authorities in Binh Thuan province) one of its tasks was to research on vine varieties for farmers in Tuy Phong district. Hence, up to now a 42 accession vine collection has been established here. Among them there are 32 table grape accessions, 5 wine grape accessions and 5 rootstocks. Today, this Centre has also released the Black Queen table grape variety for screening and pilot production on a small scale.

Research on Some Promising Table Grape Varieties

Initial results of studies on grape yield and quality in preliminary investigations showed that NH-01-48 table grape accession has more advantages than the Cardinal variety. Hence, this accession was selected for testing on a large scale. From these demonstrations carried out over several seasons, the following results were obtained:

The growth duration of NH-01-48 grape variety (from pruning to the last harvest) was about 25-30 days longer than that of the Cardinal variety. The NH-01-48 variety could resist downy mildew, bunch rot and powdery mildew better than Cardinal variety. Finally, the yield of this new variety was higher than that of the Cardinal check variety.

NH-01-48 grape variety had better quality characteristics compared with Cardinal, which included higher sugar content, fewer seeds, and sweeter taste and aroma.

Research on Micro-element Mixtures

The Cotton Research Centre succeeded in studying several kinds of mixed foliar fertilizers from macro, semi-macro and micro fertilizers, called "Duong nho". Spraying "Duong nho" on grapes increased yield and sugar content. It also increased farmers profit from VND 3 to 5.76 million per hectare (Nguyen Huu Binh, Le Quang Quyen, Vu Xuan Long, et al. 1997). This product was recognized as a technical achievement and was released for trial on a large scale by the Scientific Commission of the Agriculture and Rural Development Ministry.

Research on Growth Regulators

Based on research on the effectiveness of some growth regulators for grapes, a yield-increasing product called LH1 was found for grapes. This product contained a-Naphtyl acetic acid, gibberellic acid and some micro-elements. Berry weight, bunch weight, number of the berries per bunch and yield increased when LH1 0.1 percent was used, which especially brought profit for farmers exceeding over VND 7 million per hectare (Vu Xuan Long, Nguyen Quang Thach, 1997).

Research on Cultivation Techniques for Grapes

Currently, the Cotton Research Centre has been carrying out some research on cultivation techniques for grapes which included research on polythene film mulching, vine planting using a ditch and dyke system of land reclamation in low-lying areas, grape bunch covering techniques, use of rootstock and grafting techniques, use of economical irrigation methods, and research on some methods of agricultural extension.

CONCLUSIONS

- In Viet Nam, grapes have mainly been grown in Ninh Thuan and Binh Thuan provinces. Today, the areas under grapevines are some 2,400 hectares in Ninh Thuan (occupying 90 percent) and 250 hectares in Binh Thuan (occupying 9 percent of the total areas under grapes in the country). In the North of Viet Nam, grapes are being planted on a small scale for experimentation.
- The climatic and soil conditions of Ninh Thuan province are very suitable for the growth and development of vines. It is in this region that grapevines can be pruned at any time of the year giving 2.5-3 harvests a year.
- At present, Cardinal is an unique variety popularly grown in Viet Nam. This variety has a high yield potential but not excellent berry quality and is especially weak in pest and disease resistance. Vines are mainly propagated by cuttings.
- Ninh Thuan and Binh Thuan farmers have only used the overhead trellis system for vines. Grape planting time is mainly carried out in December and January every year. Popularly used plant density is some 2000 plants per hectare.
- The system of vine training is similar to a cobweb-like design. Pruning is done at any time of the year and irrigation is given usually every 10-15 days. Inter-cultivation is mainly done once per season together with the application of organic and inorganic fertilizers. Fertilizer doses used are rather high, most of which are given before the pruning stage.

- Vines in Ninh Thuan are infected by four kinds of diseases: downy mildew, bunch rot, powdery mildew, and rust. Among them, downy mildew, bunch rot and rust diseases often develop and cause serious damage in the wet season, but powdery mildew occurs in dry weather. In addition, vines are attacked by army worm, thrips, mealy bug and spider mites. These insects and spider mites mainly cause damage in the dry season.
- Vines in Ninh Thuan age soon because of intensive exploitation. Normally, vine growers have to get rid of old vines for replanting after about 10 years. Average grape yield in Ninh Thuan was estimated at about 30 tonnes per hectare a year. High grape yield is focused on Winter-Spring and Spring-Summer seasons but it is low and unstable in Autumn to Winter.
- There are four organizations maintaining and researching on grape collections, which are the Cotton Research Centre, Hanoi Vegetable and Fruit Research Institute, Ninh Thuan Domestic Animal and Crops Breeding Centre, and Binh Thuan Socio-Economic Development Centre. The quantities of accessions in these collections are 61, 26, 33 and 42 respectively.
- NH-01-48 table grape variety has more advantages than Cardinal variety in characters such as easy blossoming after pruning, high yield, high sugar content, delicious taste and good disease resistance.
- Because Viet Nam is a poor country, the Government has not yet been able to invest much capital for the development of viticulture. Although leaders at all levels have begun to show their concerns for viticulture development and invested on grape research, these investments are too small in comparison with the grape growers' needs in the domains of basic and applied research techniques.

GRAPE PRODUCTION IN THAILAND

Before 1960, table grapes were imported from the United States and Australia to Thailand, incurring high costs. At that time more than one hundred varieties were also introduced for testing from the United States and other countries. It was found that grapes can be adapted and grown well in the country. In 1956, Professor Pavin Punsri and his colleagues of the Department of Horticulture, Kasetsart University, Bangkok, tried to study and solve the problems of grape culture.

The grape industry in Thailand has been a success since this time. The initial commercial table grape production areas were located in the Central Plain region at Nakhon Pathom, Ratchaburi, Samut Sakhon and Samut Songkhram provinces, which are near Bangkok. At present, the grape industry has expanded to the Northern, Northeastern and the Western regions.

The climatic and soil features of the grape growing areas are described below. In the Central Plain area the range of temperature, average rainfall/year, range of relative humidity and altitude are 25-30° C, 1300-1450 mm, 60-90 percent and 2-3 m, respectively.

The soil is of a clayey type and drainage is often very poor in this region. It is, therefore, necessary to improve the drainage system by using the 'ditch and dyke system' which involves constructing raised beds alternating with ditches. The soil pH is about 5-5.5. Lime and manure are also needed to improve chemical and physical properties of these soils. The water sources in the Central Plain come from rivers and canals.

In the Northern region the range of temperature, average annual rainfall, range of relative humidity and altitude are 10-20° C, 1400-1500 mm, 60-90 percent and 1200 m, respectively. The soil is a clay loam type and well drained with a soil pH of about 4-5. The water source comes from reservoirs that have been constructed for irrigation. In the Northeastern part of the country the range of temperature, average rainfall/year, range of relative humidity and altitude are 19-30° C, 1200-1300 mm, 60-90 percent and 450 m., respectively. The soil is a clay loam type with a pH of about 3.5-5, while the water sources also come from reservoirs.

PRESENT STATUS OF GRAPE CULTIVATION IN THE COUNTRY

At present, there are about 2,717 hectares producing 31,677 tonnes/year with an average yield of about 15 t/ha. The main areas of grape production are located in the Central Plain area at Ratchaburi, Samut Sakhon and Nakhon Pathom, which mainly produce table grapes. However, some private firms use these table grapes for wine making. In the Northeastern region, growers grow both table and wine grapes. There are two wineries at Loei and Nakhon Ratchasima provinces.

In the Northern region, the growers produce table grapes in Chiang Mai and Nan provinces and wine grapes in Phichit province. In these northern parts of Thailand, grape varieties were introduced into the highlands under the auspices of the Royal Project in 1981. Research work aims to improve the living conditions of hill tribes and use grape as a substitute crop for opium poppy. Results indicated that the varieties Beauty Seedless, Ruby Seedless and Early Muscat gave good growth, production and quality. Extension activities are currently being continued.

Data on the production of table and wine grapes. The main varieties for table grapes are white Malaga and Cardinal, which produced about 27,556.7 and 4,043.9 tonnes, respectively. These two varieties are mainly used for table grapes as well as for wine making. Other less important varieties are Kyoho, Beauty Seedless, Early Muscat and Carolina Black Rose. The major varieties of wine grapes are Chenin blanc and Shiraz, which produced a premium, table white wine and a red wine, respectively.

PRODUCTION OF PLANTING MATERIAL

There are no Government nurseries to produce grape planting materials. However, the growers can buy the planting material from private nurseries, which are situated in the grape growing areas.

Grapes are commercially propagated by asexual methods. The common practices of propagating are by marcotting (air layering), marcotting followed by budding, or raising cuttings and subsequently budding onto these rooted cuttings. The methods of marcotting - budding and cutting - budding techniques are accepted by nurserymen and growers.

Solonis x othello 1613 is the only rootstock which is used for table and wine grape varieties. It is vigorous, nematode and phylloxera resistant, drought and waterlogging resistant, less resistant to lime, while at the same time it roots and grafts easily.

ESTABLISHMENT OF VINEYARDS

In the Central Plain vines are grown on raised beds (about 6 m wide) in areas that have a high watertable or are waterlogged. The beds alternate with the ditches (about 1 m wide and 1 m deep) which provide for irrigation as well as for drainage. The water is also used for mixing with fungicide and insecticide for spraying. The vines are usually grown at the centre of the bed, about 2 m apart or at the two edges of the raised beds. The number of vines per hectare varies from 550 to 1,100 vines.

In the upland areas where a high watertable does not pose a problem raised beds are not necessary. Land preparation and digging planting holes should be done after the rainy season. The size of the hole is usually 0.5 x 0.5 x 0.5 m (width x length x depth).

The spacing is about 1.5 - 2 x 3.5 - 4 m with the number of vines planted at 1,250-1,875 density per hectare. Soil in the planting holes is allowed to dry for one month before planting. Manure or compost should be mixed with the soil and filled up in each hole. The suitable schedule for planting is in January or February.

CARE AND MANAGEMENT OF VINEYARDS

Support

Grapevines cannot be grown satisfactorily without some support. There were many kinds of training systems during the early years of grape culture in Thailand. At the present, the arbor system is preferred to a trellis. The height of the arbor is about 1.5 m above the ground level. The top is about 2-3 m wide.

However, the training systems are dependent on the varieties of grapes, climate, and objectives of the growers.

Training and Pruning

In training a vine, the canes are spread out evenly to cover the area of the arbor, with no definite or regular training design. When the vine is 10 to 12 months old, the mature canes are pruned in order to force out the buds.

Fertilizer Application

Grapevines can adapt themselves to a wide range of soil fertility conditions. However, fertilizers are necessary for vigorous growth, high yield and to give better quality fruits. The main areas of grape culture in Thailand have clay soil types and were formerly rice fields. These soils need a large amount of organic matter and, therefore, manure is used a great deal. Complete fertilizers are usually applied to grapevines in the following manner:

- During the early growth phase of young vines obtained by budding, grafting or marcotting or before the first pruning, ammonium sulphate, urea or complete fertilizer (15:15:15) at the rate of 200-300 g/plant are usually applied every month.
- One to two months before the regular pruning period, fertilizer (9:25:25) at the rate of 200-300 g/plant is applied to promote flower bud development.
- Two weeks after fruit set, complete fertilizer (15:15:15) at the rate of 200-300 g/plant is applied to facilitate fruit growth and development.
- During the beginning of fruit softening or during colour change about one month before harvesting, complete fertilizer (13:13:21 or 12:12:17:2 or 9:25:25) at the rate of 200-300 g/plant should be applied to enhance fruit quality.

Weeding

In the Central Plain area, weeds are usually controlled by hand hoeing and application of herbicides. The herbicide used is grammoxone, which destroys the green parts of the weeds as the farmers prefer clean cultivation. In the uplands, weeds are controlled by hand hoeing, ploughing and using herbicides.

Irrigation

Water source and supply are important for the vineyards. Reservoir and groundwater tube wells are suitable for surface irrigation. Pipeline systems and hose irrigation would be practical and more economical for the farmers. However, mulching with rice straw is also very necessary during the dry season. In the Central Plain areas, the vines are grown on raised beds alternating with the ditches, which can provide the water supply needed for the vines.

Insects and Diseases

Insect problems are not that serious but diseases such as downy mildew, powdery mildew, anthracnose, bitter rot and fruit rot are quite harmful.

Insects: There are several kinds of insects that attack leaves, flower and berries e.g. leaf hopper, leaf roller, thrips and mites. The insecticides used for control are carbaryl, methomyl, carbosulfan, and dicofol, etc.

Diseases: The important diseases of grapes are described below:

a) Downy Mildew: The fungus (*Plasmopara viticola*) thrives best and becomes most destructive during still, cool and moist weather. Although primarily a foliage disease, the fungus also attacks flower clusters. The fungicides used for control of downy mildew are Zineb, Maneb, Captan, Mancozeb and Metalaxyl.
b) Powdery Mildew: The causal fungus (*Oidium tuckeri*) unlike most other diseases of grapes, which are favoured by moist conditions, thrives best in a dry climate. It attacks during berry growth and these developing berries will crack later. For the control of powdery mildew, the vines are sprayed with wettable sulphur.
c) Anthracnose: The causal fungus (*Gloeosporium ampelophagum*) attacks all parts of vines. The fungal attack is usually serious during the rainy season. For the control of anthracnose, the vines are sprayed with Mancozeb, carbendazim or benomyl.
d) Dead Arm: The fungus (*Melanconium fuligeneum*) attacks during fruit maturity, particularly during 2-3 weeks before harvesting. The affected canes will die and the fruit will soften and shrink. For the control of this fungus a spraying of copper fungicide or difenoconazole is given to all parts of the vines.
e) Berry Rots: There are several fungi which will attack the berries and fruit clusters during fruit maturity such as Blue mold (*Penicillium* spp.), Rhizopus rot (*Rhizopus* spp.), Ripe rot (*Collectotrichum gloeosporioides*), Bitter rot (*Greeneria uvicola*) and Botrytis bunch rot (*Botrytis cinerea*). The fungal attack is usually serious during the rainy season. The control of these fungi is by spraying common systemic fungicides.

Special Practices for Improving Grape Quality

There are some special practices to improve the yield and quality of grapes such as:

- Improving bud break or bud burst is achieved by treating canes with hydrogen cyanamide.
- Cluster elongation, cluster loosening, berry enlargement and seedlessness are induced by treating with gibberellic acid (GA).
- Increasing fruit colouration can be achieved by ethrel (2-chloroethylphosphonic acid) application.

- Rain and bird damage can be controlled by using a plastic roof and bagging of bunches.

HARVESTING AND YIELDS

Thailand has a tropical climate, which seems to stimulate the growth of grapes immensely. The first crop can be expected only 14 to 16 months after planting. The harvest can be timed at will because planting and pruning can be done at anytime of the year.

The buds of the canes can be forced to sprout at any time of the year by pruning and consequently, two to three crops can be harvested from an individual vine in a year.

In practice, however, the growers prune their vines twice a year and get two crops, one in the rainy season (May to October) and the other crop in the drought season (November to April). The latter crop is superior due to higher sugar content and better appearance. In the Central Plain area, grapes are grown under the ditch and dyke system. The pruning can also be done at any time of the year. However, in this intensive culture system disadvantages can be expected as the vines have a short life span of only 7 to 10 years. In general, the crop is inferior in quality during the rainy season and vineyard management is also quite intensive.

In White Malaga, which is a late variety, growers can have two harvests a year per vine or five harvests per vine every two years. The average yield is 15-20 kg per harvest. With Cardinal, a very early variety, one can have even three to four harvests a year per vine. The average yield of a vine is 12-15 kg per harvest.

MARKETING

Marketing of grapes in Thailand is not organized as there is no organization or association for marketing of this crop. Hence, the sequence of marketing is usually from the growers to middlemen (1 or 2) who transport to markets from which the produce is retailed to consumers. Most of table grapes are for domestic consumption. The export of table grapes is only a small amount. The price of table grapes depends on the variety, quality, season and demand. The average farm price of old varieties like white Malaga and Cardinal is about 30-37 baht/kg, while that of the new variety of seedless grapes is about 100-150 baht/kg.

Table. Quantity and Value of Grape Exports (1993-1997)

Item	1993		1994		1995		1996		1997	
	Quantity	Value	Quantity	Value	Quantity	Value	Quantity	Value	Quantity	Value
Grapes Fresh	40	861	34	707	31	545	22	509	38	839

The quantity and value of table grapes and wine imports. The quantity and value of table grapes and wine have increased since 1993 up to the present time.

Table. Quantity and Value of Grape Imports (1993-1997)

Items	1993		1994		1995		1996		1997	
	Quantity	Value	Quantity	Value	Quantity	Value	Quantity	Value	Quantity	Value
Grape Fresh	1,633	131.7	2,326	177.4	2,318	183.4	2,591	191.6	3,640	262.7
Grape Dried	81	4.9	53	2.5	79	3.5	189	6.9	206	9.3
Total	1,714	136.6	2,379	179,9	2,397	186.9	2,780	198.5	3,846	272

Table. Quantity and Value of Wine Imports (1993-1997)

Items	1993		1994		1995		1996		1997	
	Quantity	Value	Quantity	Value	Quantity	Value	Quantity	Value	Quantity	Value
Champagne (1000 litres)	157	30.9	189	25.92	396	39.6	89	21.5	76	18.0
Wines (1000 litres)	111	10.1	114	11.8	193	40.7	258	42.8	123	10.3
Other wines (1000 litres)	1,824	141.5	1,907	169.9	3,657	360.8	8,579	784.8	5,734	410.6
Total	2,092	182.5	2,210	207.6	4,246	441.1	8,926	849.1	5,933	438.9

POTENTIAL FOR GRAPE PRODUCTION DEVELOPMENT

In the past, table grapes were imported from foreign countries at great cost. Since 1960, grape growers have succeeded in producing grapes locally and even have exported to foreign markets. In 1987, however, new varieties of good quality table grapes continued to be imported in increased quantity and value although the imported table grapes were expensive. Both table and wine grapes have a very good potential as a horticultural fruit crop, although it requires an intensive crop management. Nevertheless, it is a lucrative enterprise for farmers.

CONSTRAINTS IN GRAPE PRODUCTION DEVELOPMENT

Thailand is located within the tropical climatic belt characterized by a hot climate, high rainfall and humidity. However, the vines can be adapted to grow well. Although the fruit produced in the dry season is of very good quality, the production, productivity and quality of grapes is far below that of sub-tropical climatic regions. Grape production can be useful for local consumption and export to some countries in the Region. The advantage of grape production in the country is that growers can predict the harvesting time of the year by pruning, and the vines can be harvested two to three times a year. There are several constraints to be considered that affects the development of the grape industry in Thailand.

- In the Central Plain or lowland areas, the growers have to construct raised beds or dykes alternating with ditches. The growers allow too much over-cropping, often harvesting two-three times a year. The vines, therefore, have a short life-span of about 7-10 years under this intensive cropping system.
- The temperature, rainfall and high humidity are limiting factors that affect produce quality.
- The growers spray a lot of fungicides and insecticides to protect their crops without considering residual effects of the chemicals. The growers learn their grape cultivation from nearby grape growers and they also use these potent chemicals on the advice of the chemical suppliers.
- High initial investment, especially high cost of new good varieties, trellising, constructing reservoirs and irrigation systems, agricultural tools, framework for plastic roofs to protect from the rain, high cost of insecticides, fungicides and fertilizers are constraints to the expansion of the grape industry.
- Lack of knowledge on diseases and pest control measures and quality control technologies.
- Inadequacy of sufficient Government policy support for the grape industry.

GOVERNMENT POLICIES AND PLANS FOR RESEARCH AND DEVELOPMENT OF GRAPES

The priority of the important economic fruit crops are considered by the Committee of the Horticulture Research Institute under the Department of Agriculture, Ministry of Agriculture and Cooperatives. The grape crop has been given the lowest priority among fruit crops. There are some plans for wine and table grape production development in 1977-2001 and the period of the next ten years beyond 2001. The aim is to study all aspects of production. In the past, there were only a few research papers on grapes. However, the Department of Horticulture, Faculty of Agriculture, Kasetsart University, has tried to study and carry out research on table and wine grapes since 1960 up to the present time.

CONCLUDING REMARKS

The grape industry in Thailand has progressed greatly since 1960. The main areas are located in the Central Plain, which is diversifying into the rice growing lands. The major grape varieties are White Malaga and Cardinal, which are used for table grape and wine making. At present, the grape industry has expanded to Northern, Northeastern and the Western regions. The total planted area of 2,717 hectares produces 31,677 tonnes/year and the average yield is about 15 t/ha. Of the total grape production, 87.4 percent is used as table grapes and 12.6 percent for wine production, respectively. Vineyard management is quite intensive and often costly to growers. There are many diseases and insect problems. Grape growing, however, is a highly remunerative enterprise as the crop is easily marketable. The grape industry development should be supported by the Government through the cooperation of the researchers and the growers. Considering the increasing demand and the ingenuity of local farmers, the grape industry has much potential for further expansion in the years to come.

8

Method of Diseases Control

INTRODUCTION

By the time disease symptoms appear, disease pathogens are inside the plant and usually beyond control. Therefore, it is important to prevent penetration of pathogens. Consider the following methods of disease control:

Avoid

Avoid certain diseases through choice of appropriate site and planting time, purchase of disease-free stock and cultural practices that do not favour disease infection. Insect control can be important for controlling spread of viruses, mycoplasmas, bacteria and fungi. Mulching can help control diseases by preventing the contact of foliage with soil.

Exclude

Exclude insect vectors with row covers. Don't bring diseased plants into the garden.

Use Resistant Or Tolerant Plants

Appropriate selection of species or cultivars will avoid or minimize problems. Remove all diseased plants, including alternate hosts. Immediate removal of diseased plants or plant parts reduces the chance of the disease spreading. Selective pruning and careful sanitizing of pruning equipment can prevent spread of disease. Rotate crops to avoid soil-borne diseases. Use best cultural practices. Follow recommended cultural practices.

TREAT

Treat plants with recommended chemicals to prevent infection. Some treatments will stop the spread of the pathogen.

PEST-RESISTANT CROPS

One of the mainstays of integrated pest management is the use of crop varieties that are resistant or tolerant to insect pests and diseases. A resistant variety may be less preferred by the insect pest, adversely affect its normal development and survival, or the plant may tolerate the damage without an economic loss in yield or quality.

Disease-resistant vegetables are widely used, whereas insect-resistant varieties are less common but nonetheless important. Examples include varieties of wheat which have tough stems that prevent development of the Hessian fly and cucurbits (squash, cucumbers, melons) that have lower concentrations of feeding stimulants (cucurbitacins) for cucumber beetles.

Commercial corn lines with increased concentration of a chemical called DIMBOA display resistance to European corn borer, a major pest throughout the world. In the case of cabbage, the only reliable method of controlling onion thrips is through the use of resistant varieties.

Advantages of this tactic include ease of use, compatibility with other integrated pest management tactics, low cost, and cumulative impact on the pest (each subsequent generation of the pest is further reduced) with minimal environmental impact.

The development of resistant or pest-tolerant plant varieties, however, may require considerable time and money, and resistance is not necessarily permanent. Just as insect populations have developed resistance to insecticides, populations of insects have developed that are now able to damage plant varieties that were previously resistant.

CHEMICAL CONTROL

If all other integrated pest management tactics are unable to keep an insect pest population below an economic threshold, then use of an insecticide to control the pest and prevent economic loss is justified. In most cropping systems, insecticides are still the principal means of controlling pests once the economic threshold has been reached.

They can be relatively cheap and are easy to apply, fast-acting, and in most instances can be relied on to control the pest(s). Because insecticides can be formulated as liquids, powders, aerosols, dusts, granules, baits, and slow-release forms, they are very versatile.

Insecticides are classified in several ways, and it is important to be familiar with these classifications so that the choice of an insecticide is based on more than simply how well it controls the pest. When classified by method of application (site of encounter by insect), insecticides are referred to as stomach poisons (those that must be ingested), contact poisons, or fumigants. The most precise method of classifying insecticides is by their active ingredient (toxicant).

According to this method the major classes of insecticides are the organophosphates, chlorinated hydrocarbons, carbamates, and pyrethroids. Others in this classification system include the biologicals (or microbials), botanicals, oils, and fumigants.

Insecticides may be divided into two broad categories: (a) conventional or chemical and (b) biorational. In this guide we define conventional or chemical insecticides as those having a broad spectrum of activity and being more detrimental to natural enemies.

In contrast, insecticides that are more selective because they are most effective against insects with certain feeding habits, at certain life stages, or within certain taxonomic groups, are referred to as "biorational" pesticides. These are also known as "least toxic" pesticides.

Because the biorationals are generally less toxic and more selective, they are generally less harmful to natural enemies and the environment. Biorational insecticides include the microbial-based insecticides such as the Bacillus thuringiensis products, chemicals such as pheromones that modify insect behaviour, insect growth regulators, and insecticidal soaps.

The majority of insecticides fall into the chemical category because they are typically more effective and can usually be used to control several pests. This provides the economic justification needed for the research and development of such products. In contrast, the more specialized market of the biorationals makes their long-term economic return less favourable.

Despite the advantages of conventional insecticides, the problems associated with their use have been well documented. These include the resurgence of pest populations after decimation of the natural enemies, development of insecticide-resistant populations, and negative impacts on nontarget organisms within and outside the crop system. One of the more serious problems is the development of resistance. Many insect pest species now possess resistance to some or several types of insecticides, and few chemical control options exist for these pests.

If all other integrated pest management tactics are unable to keep an insect pest population below an economic threshold, then use of an insecticide to control the pest and prevent economic loss is justified. In most cropping systems, insecticides are still the principal means of controlling pests once the economic threshold has been reached.

They can be relatively cheap and are easy to apply, fast-acting, and in most instances can be relied on to control the pest(s). Because insecticides can be formulated as liquids, powders, aerosols, dusts, granules, baits, and slow-release forms, they are very versatile.

To compare pesticide use patterns, the dose, formulation or per cent active ingredient of the product, and the frequency of application must also be considered and are used to calculate the EIQ Field Use Rating.

Pesticides with a lower EIQ Field Use Rating have a lower environmental impact. The EIQ concept and techniques are evolving, but this, or a smilar rating, will provide information for appropriate pest management decisions in the future.

Natural enemies are generally more adversely affected by chemical insecticides than the target pest. Because predators and parasitoids must search for their prey, they generally are very mobile and spend a considerable amount of time moving across plant tissue. This increases the likelihood that they will contact the insecticide. When an insecticide is applied, ideally only the target pest(s) should be affected. The goal is to maximize pest mortality while minimizing harm to natural enemies.

The following factors, some of which are used when determining the EIQ, influence the impact of insecticide applications on natural enemies:

- Spectrum of activity. Insecticides toxic to most insects (broad-spectrum) will more adversely affect natural enemies than materials that are narrow-spectrum or selective for specific insect species or life stages. Most insecticides in use today have a broad spectrum of activity.
- Residual (half-life) activity of insecticide. Insecticides that remain toxic to pests for a long time and remain on the treated surface will have a similar effect on natural enemies.
- Coverage and formulation of insecticide. Full coverage sprays will generally have a greater impact on natural enemies than directed sprays, systemics, or bait formulations. Spot or edge treatments directed at localized pest infestations or to a specific plant surface most often occupied by the pest will be less detrimental than those applied to the entire field or plant.
- Dosage and frequency of application. Higher rates and repeated applications will have a more detrimental impact on natural enemies.
- Timing of application. Applying insecticides when natural enemies are not abundant or are less susceptible, such as when immatures are encased in host eggs, can be helpful.
- Susceptibility of natural enemy to insecticide. Some natural enemies are inherently more resistant to insecticides than others, and some populations of natural enemies have been selected either naturally or through the efforts of researchers to possess higher levels of resistance.
- For example, strains of predaceous mites, rove beetles, lacewings, and species of parasitoids have been selected for increased levels of resistance to insecticides, thereby increasing the likelihood that some will survive insecticide treatments directed at pests.

  Some of these pesticide-tolerant strains are commercially available. Applications of some fungicides directed at controlling plant diseases can also reduce the incidence of fungal diseases of insects. In general, little can be done about this incompatibility except to use the fungicide only as needed.

In addition to killing natural enemies directly, insecticides may also have sublethal effects on insect behaviour, reproductive capabilities, egg hatch, rate of development, feeding rate, and life span. Fungicides and herbicides may also have lethal and sublethal effects on natural enemies.

CULTURAL CONTROL

There are many agricultural practices that make the environment less favourable to insect pests. Examples include cultivation of alternate hosts (e.g., weeds), crop rotation, selection of planting sites, trap crops, and adjusting the timing of planting or harvest.

Crop rotation, for example, is highly recommended for management of Colorado potato beetle. The beetles overwinter in or near potato fields and they require potato or related plants for food when they emerge in the spring.

With cool temperatures and no suitable food, the beetles will only crawl and be unable to fly. Planting potatoes well away from the previous year's crop prevents access to needed food and the beetles will starve. The severity and incidence of many plant diseases can also be minimized by crop rotation, and selection of the planting site may affect the severity of insect infestations.

Trap crops are planted to attract and hold pest insects where they can be managed more efficiently and prevent or reduce their movement onto valuable crops. Early planted potatoes can act as a trap crop for Colorado potato beetles emerging in the spring. Since the early potatoes are the only food source available, the beetles will congregate on these plants where they can more easily be controlled.

Adjusting the timing of planting or harvest is another cultural control technique. The earlier planted processing tomatoes grown in the western United States are far less likely to be infested by the tomato fruitworm than those planted later in the season. It is also important to use pest-free transplants. Some vegetable crop transplants can be infested with insect pests, and growers using these transplants are put at a considerable disadvantage.

CULTURAL METHODS OF DISEASE CONTROL

Stevens discussed the cultural methods of disease control. According to Stevens, these methods involve agricultural cropping, harvesting and storage, tillage, crop rotation, soil management, growing of resistant varieties, planning of land use, and other practices.

1. Avoidance of the Pathogen

Many plant diseases can be totally avoided by a properly selecting of the land or field, time of sowing, selection of varieties, seed and planting stock and by improving of cultural practices. The aim of these measures is to enable the host to avoid contact with the pathogen or the susceptible stage of the plant and the favourable conditions for the pathogen should not coincide.

a. Selection of Sowing Area

Many fungal and bacterial diseases can be seen more in wet areas than in dry areas. The crops which are prove to these diseases, if grown in wet areas, are very likely to be affected by plant pathogens. *Tolyposporium penicillariae,* the pathogen responsible for the smut disease of bajra and the ergot of bajra caused by *Claviceps microcephala,* are more severe in wet areas in regions where rain occurs for long durations during the flowering stage of the crop.

b. Selection of Field

The selection of a suitable area or field for growing of crops is very important from the point of view of better results as well as protection of the crop from the clutches of pathogens. In the case of many soil borne pathogens, if the same field is chosen for cultivating a specific crop, there is every chances of the disease appearing in severe form due to the build up of the inoculum potential of the pathogen. Hence, it is advised not to grow the same crop in the same field, where in the previous year, it was infected with a pathogen capable of surviving in the soil. The examples are wilt of arhar, green ear disease of bajra, stem, powdery mildew of wheat, smut of bajra, ergot of bajra, ear cockle of wheat and bacterial wilt of solanaceous -plants.

The drainage conditions of the field are also important. Low lying, waterlogged fields will produces such diseases as red rot of sugarcane and downy mildew of bajra.

In the case of fruit orchards, the selection of a suitable field is very important. If apple trees are planted in areas where oak trees were previously grown, there are chances of collar rot due to *Rosellinia* sp. becoming severe.

c. Time of Sowing

Pathogens are able to attack susceptible plants only under certain favourable environmental conditions, For example, downy mildews require continuous high water content for infection. If by some means, the crop is grown in such. a way that it is not attacked by pathogens during its most susceptible period, disease incidence can be considerably reduced. *Rhizoctonia* root rot of gram is severe if gram is sown immediately after the rains. This is due to the fact that the pathogen plant develops rapidly under higher temperature and moisture conditions, A. slight delay in sowing is helpful in this case.

d. Disease Escaping Varieties

Many varieties of crops prevent themselves from damage because of their growth characters. These varieties escape the attack of the pathogen and do not resist the attack due to their inherent characteristics, for example, early maturing varieties of wheat or pea escape damage due to *Puccinia graminis trifid* and *Erysiplze polygoni* respectively.

e. Selection of Seed and Planting Stock

In India seed potatoes are generally imported in the southern states from the Simla hills for better control of viral and bacterial diseases as in simla hills disease resistant potatoes are researched and produced for all Inida. In the USA, seed growing areas have been shifted to dry pacific regions for crops such as cabbage, turnip, beans and peas for obtaining disease-free seed and controlling such diseases as black leg and black rot of cabbage and turnip. Since many plants grow by vegetative parts, the selection of disease-free planting material forms a very important control measure.

The sowing of disease-free seeds in a disease-free field is often a very important control measure for certain diseases, such as red rot of sugar cane, black scurf of potato, or various viral diseases of potato. Tuber indexing has been resorted to for obtaining virus-free seed and protecting crops against the attack of various viral diseases of potato.

f. *Modification of Old Cultural Practices*

The distance between plants in the field time and amount of irrigation, quantity and quality of fertilizers or organic manures, time and method of planting, mixed cropping method, depth of sowing, etc. are some of the cultural practices which effects the seriousness and occurrence of disease of certain diseases. The changes needed in these practices can help to reduce disease incidence and loss of the crop.

Tillage provides a temporary improvement in aeration, accelerates drying and redistributes components in the soil. These environmental and physical changes can predispose pathogens to antagonistic effects of the microbiota. Early and deep ploughing kills cotton roots and favours saprophytes rather than *Phymatotrichum omnivorum*. According to Garren and Duke, if the plant debris is buried deeper by ploughing, powered the pathogen becomes less potent.

Wrong mechanical sowing makes plants prone to infection. This may also result in an unequal and uneven distribution of seed. Overcrowding of seeds and plants may result in damping off of seedlings. The depth of sowing of seeds has an important effect on some pathogens, notably those which attack seedlings. By delaying the emergence of seedlings, deep sowing may help to increase the resistance of a prove crop to a pathogen. Good control of blind seed is obtained by planting infected seed at least 1.25 cm deep and with complete soil coverage. Mixed cropping is of special significance in Asian countries. This practice helps to effectively check the spread of infectious diseases from one crop to another

Sclerotium rolfsii on pea-nuts is controlled by ploughing under the surface refuse before sowing. This deprives the fungus of a food base since it fungus grow only on or near the soil surface. The effect, therefore, is very much like that of crop rotation. Deep ploughing of soils that contain diseased potato tubers can effectively decrease the amount of inoculum or *Phytophthora infestans.* Mixed cultivation of cotton and moth *(Phaseolus aconitifolius)* reduces the incidence of root rot of *the* former crop. Mixed crops, such as wheat and barley, wheat and gram, arhar and jowar, cotton and moth reduce the economic losses from diseases.

Management of soil water offers a means to control the associated micro biota-directly, through water potential, and indirectly, through gas exchange of the soil. Common scab of potato can be managed by keeping the soil water potential near the tubers higher than -0.4 bar during tuber formation . The dry land foot rot of wheat caused by *Fusarium roseum* 'Culm arum' car be controlled by irrigation, partly because the water reduces stress in the host (Papendick and Cook, 1974), and also because the germlings of *Fusarium* lyse or convert into new chlamydospores at water potentials above -10 to -15 bars.

In Israel, the diseases of tomato caused by *Stemphylium botryosum* and *Xanthomonas vesicatoria* were increasd by overhead sprinkling, whereas *Leveillula taurica,* a powdery mildew which thrived in dry conditions, was reduced. In the presence of excess free water, bacterial cells and fungal spores are disseminated easily.

Pathogens, such as *pythium* and *Phytophthora* species, are dependent on high soil moisture content. Cereal rusts are more common in heavily irrigated crops or those grow in very wet soil.

2. Exclusion of Inoculum

The main aim of these method is prevent pathogens from infecting virgin area and crop of that area. Seed certification, crop inspection, growing crops in regions unfavourable for pathogen and quarantine measures are some of the measures of preventing the spread of pathogens.

a. *Seed Treatment*

Seeds, tubers, grafts, bulbs and other propagating materials can be given heat, gas or chemical treatment to keep the seeds free of pathogens.

b. *Quarantine Regulation*

A quarantine can be defined as a legal obtainment on the movement of agricultural products for the purpose of exclusion, prevention or delay in the spread of plant pests and diseases in uninfected areas. In most agricultutally advanced countries, plant quarantine rules and regitation has been placed on the statute book to restrict movement of diseased plant material or of fungi, bacteria or viruses that cause diseases in plants. Quarantine measures are of three types:

(a) domestic,

(b) internal, and

(c) total embargoes.

The quarantine laws were first passed in the USA, in 1912 and are known as Federal Quarantine Acts. In India, the Destructive Insect and Pest Act was passed in 1914 and subsequently supplemented by other legislations.

The DIP Act in India was passed in 1914 and has been revised eight times from 1930 to 1956 and corrected up to 1967 but it still is not enough to fulfil the needs under the prevailing circumstances. In India there are 16 quarantine stations operating under the Directorate of . Plant Protection and Quarantine—eight at sea ports, six at airport and two on land frontiers (Hussainiwala in Ferozepur and Sukhiapokri in Darjeeling district). There are a number of gaps in the DIP Act, and as a result many serious plant pathogens have already been introduced in the country. These are now heavily destroying crops in certain areas. Golden nematode and wart diseases of potato have been introduced in India from the European countries and are now well established in the Nilgiri Hills and Darjeeling. Fortunately, timely action has prevented the spread of wart disease to other parts of India.

In India, so far, quarantine acts and regulation exist for two insect pests, i.e. Fluted Scale *(Tcerva purcheri)* and San Jose Scale *(Quadraspidiotus permiciosus)* and for three diseases, i.e. wart *(Synchytrium endobioticum)* of potato, bunchy top (virus) and mosaic (virus) of banana domestic with a view to prevent the spread of these pests.

The problem of plant disease is international. Hence, in 1951 in Rome, an International Plant Protection Convention was drawn up which at present has about 50 signatory nation members. Briefly, each contracting government agrees to make provisions for:

- An official protection organization with the specific basis of inspecting, growing crops and the produce derived from them and issuing phytosanitary certificates.
- The distribution of information regarding pests arid diseases both within the country and to FAO so that a world reporting service is established.
- Research and investigation in the field of plant protection on a cooperative basis for disease which had international effects. Within the framework of the international cooperation there are six regional groups.

a. European Plant Protection Organization (EPPO). This was formed prior to the treaty in Rome.
b. Inter African Phytosanitary Commission (for South Africa, South of Sahara).
c. Plant Protection Committee for South East Asia and the Pacific region.
d. Organizmo International Regional de Sanidad Agroperciane (Central America and Mexico).
e. Convenio Intermericano de Protection Agricola (South America).
f. Near East Plant Protection Commission.

3. Eradication of the Pathogen

The main aim of these measures is to break "infection chain" by removing the focus infection and starving the pathogen. A knowledge of the life cycles of the pathogens; method of perennation, host range and habit of growth are essential prerequisites for introducing suitable measures to prevent the disease. Eradication involves totally removing the pathogen from infested areas, and the magnitude of the operation involved may differ considerably. Eradication is attempted through roguing, sanitation, crop rotation, eradication of alternate and collateral hosts, heat and chemical treatment of diseased plants, soil treatments, biological control, etc.

a. *Roguing or total destruction of infected plants*

This practice contains careful removal and destruction of infected plants or plant parts - from the field at an early stage by removing the "foci" of infection and preventing wide spread of the pathogen. Rouging is employed in such diseases as loose smut of wheat, loose and covered smut of barely, sorghum, maize, red rot of sugarcane, ergot of bajra, wilt of arhar, ,and many viral diseases.

b. *Eradication of Alternate or Collateral Hosts*

Many plant pathogens needs two hosts to complete their difference; one of them is generally a wild plant. The destruction of such a wild host plant breaks the chain in the completion of the life cycle of the pathogen. Many plant pathogens have several alternate hosts on which they pass their life cycles. The destruction and removal of collateral hosts is an important control measure of the disease. The yellow vein, mosaic of lady's finger persists on wild hosts *(Hibiscus tetraphyllus)* in nature and the systematic removal of this wild plant helps to control the disease.

c. *Crop Rotation*

Very often, regular cultivation of the same crop or related crops in the same field leads to the perpetuation of a pathogenic soil-borne fungus and the gradual increase in the intensity of the disease. Soils generally suffer from what we call soil sickness. When the same crop is raised year after year in the same field, the soil-borne pathogens of that crop easily establish themselves in the soil and increase their population.

After a certain period of time, the soil becomes so heavily populated with the pathogen that it no longer supports the growth of the crop. Such soils are called "sick soils". Crop rotation helps to 'control many soil-borne diseases, such as wheat mosaic, wilt of linseed; finger and toe disease of crucifers, ergot and smut of bajra, bunt of wheat, leaf smut and bunt of rice.

Curl has studied & reviewed the control of plant diseases by crop rotation. This method is also important with other point of view. If the same crop is sown in the same area year after year, there are chances that a particular essential nutrient in the soil will be exhausted.

There can also be the accumulation of toxic substances and organic acids if a specific crop is continually cultivated in the same area. Crop rotation is generally recommended which require totally different nutrients and affected by different pathogen.

d. *Sanitation*

Like plant sanitation, field sanitation is also essential for the control of many plant diseases. Pathogens survive through dormant structures in plant organs in the soil. The collection and destruction of these structures is an important control measure. It is common practice to bum diseased stubble for plant disease control.

The effectiveness of fire for controlling plant disease has been reviewed by Hardison. In fact, sanitary conditions are helpful in almost all infectious diseases. The spread of certain diseases, for example, tomato mosaic virus, would not be rapid if the workers who smoke were to wash their hands before handling the plants.

e. *Heat or Chemical Treatment of Diseased Plants*

The pathogen present in the plant can be killed or deactivated by applying heat or chemicals. Hot water treatments have generally been most successful in controlling diseases and are used to eliminate *Ustilago nuda* from wheat and barley seed, the nematode *Dictylenchus dipsaci* from *Narcissus* bulbs and, ratoon stunt and other diseases from sugar cane setts.

Treating the seeds with fungicides, especially organ- omercurials, is new varieties of a common practice to remove the superficially borne inoculum, and there are now systemic fungicides available which can eliminate deep seated fungal infection. More recently, thiram soak has been successfully utilized to eliminate *Septaria appiicola* from celery seed, and *Mycosphaerello pinodes* and *Ascochyta pisi* from pea seed.

f. *Soil Treatments*

Method of soil treatment involves the use of chemicals, heat flooding, etc. Plant pathogens can be killed & removed from the soil by dry heat, steam or hot water, flooding and fumigation with chemicals. Heat and chemical treatments have been successfully used for the eradication of plant pathogens from green house soils as well as soils in plant beds. The treatment of soils in large fields has not been so effective. Nematodes can be killed in the soil by fumigating it with DD mixture or with other nematicide. PCND has been found to be effective in eradicating sclerotia-producing soil pathogens, formaldehyde in controlling onion smut disease.

g. *Biological Eradication of Plant Pathogens*

The term biological control clearly means control of a disease through some biological agency, and the term 'biological agency', means a living micro-organism or macro-organism other than the diseased or damaged plant acting as host and the pathogen or pest causing the disease or damage. According to Garrett, "Biological control of plant disease may be defined as any condition or method hereby survival or activity of a pathogen is reduced through the agency of any other living organism (except man himself), with the result that there is a reduction in the incidence on he disease caused by the pathogen".

In 1951 a new method was approached by Bliss in the USA who was working on root disease of citrus caused by *Armillaria mellea.* He suggested that if the soil is fumigated with carbon disulphide, the disease producing fungus is killed not directly by the fumigant but indirectly by the produce and dominance of *T. viride,* the population of which gets increased after fumigation. Garrett tested the effect of fumigation and the antibiotic effect of *T viridearid* assessed the relative parts played by these two components. He found that about 30% of *Armillaria mellea* was killed due to the dominance of *T. viride,* while the rest was directly killed by the fumigant. He also pointed out that in the thicker plant roots, *T. viride* may be more effective in the process of killing than the fumigant which might not be able to reach easily.

The most conspicuously successful biological control is by Rishbeth in England. Rishbetb has been studying a root disease of pines *(Pinus sylvestris* and *P. nigra)* caused by basidiomycetous fungus *Fomes annosus.* The disease spreads through infection by basidiospores produced in the large fructifications. The spores infect the left-over stumps when the trees are felled. Rishbeth at first tried to control.

The disease by painting these stumps with fungicidal paints, such as creosote. This brought only partial control because the stump tissue cracked with progressive drying, causing fissures through which the fungal spores of *F. annosus* could enter and cause infections. Rishbeth's research associate, Meredith made an important observation that along with *Fomes annosus* there were a few other fungi which were of the nature of weak parasites that colonized the cut surfaces of stumps. Rishbeth has pointed out that he originally selected *Peniophora gigantea* for this purpose because it is the most common and effective competitor of *F. annosus* in the natural colonization of pine stump surfaces. It decomposes lignin and cellulose vigorously and the fungus establishes itself in the stump quickly and from there into the stump roots.

From these stump root infections the pathogen invades the roots of other healthy trees. Rishbet showed that if paste of oidia of *P. gigantea is* pointed on the stump, it capture it very tart and kills the tissues, making it easy for saprophytes to colonize and decompose the stump and making the tissues totally unsuitable for an attack by *Fomes annosus* which is a specialized parasite. Rishbeth determined that a dose of 1 × 10" oidia of *Peniophora gigantea* allowed a satisfactory margin of safety for effective biological control of the largest spore inoculum of *F. annostls* for a pine stump of 16 cm wood diameter.

h. Soil Amendments

Many root diseases have been controlled by ploughing organic matter into the soil, such Several others have been successfully controlled in this way, such as *Phymatotrichum* root rot of cotton, "take all" *(Ophiobolus graminis),* several *Rhizoctonia* and *Fusarium* diseases.

In the case of *Ophiobolus graminis,* it has been shown that organic amendments, rich in carbon and deficient in nitrogen, control the "take all" disease of wheat. This can be explained as follows: As a result of organic amendment with a high C/N ratio, there is a big increment in the microbial number and activity. There is considerable liberation of CO_2 by soil saprophytes which diminishes the pathogenic activity of this fungus. In the process of survival also, low nitrogen content in the soil reduces the life of the fungus.

The organic amendments increase the activity of saprophytic organisms in a flush which results in the abundant production of CO_2 as a result of the respiratory activity of the micro-organisms. The pathogens sensitive to CO_2 are killed. Due to the rapid multiplication of micro-organisms in the soil, the available nitrogen in the soil is rapidly used by the fast growing saprophytes. This results in acute nitrogen shortage which adversely affects the growth of the pathogens.

In their studies on the control of bean root rot with barley amendments, Snyder established that additions of nitrogen nullified control. The germination of the chlamydospores is poor and the hyphae produced are also depauperate. This is perhaps due to the locking up of avail able soil nitrogen in the cells of micro-organisms which decompose the organic material. This is thought to deprive the chlamydospores of the pathogen of even the minimal amount to soluble nitrogen required for germination and host, infection. The pathogen and the disease develop rapidly, if nitrogen is increased. It has been shown that the inability of the conidia to germinate & grow in aqueous extracts of soil solution with cellulose (C/N ratio higher than 2:1) and barley (C/N ratio 55) can be overcome by as small a quantity of nitrogen as 5 ppm. All these factors show that nitrogen limits pathogenesis.

Extensive data about the behaviour of *Rhizoctonia solani* have been provided by Papavizas and Davey. Davey and Papavizas for the control of *Rhizoctonia* disease of snap bean *(Phaseolvs vulgaris)* by the use of organic amendments. Blair in studies with Rossy Chlodny slides demonstrated that the growth of *Rhizoctonia solalli* in unsterilized soil is checked by the incorporation of *1%* finely divided organic matter, such as dried grass, alfalfa or wheat straw; Concerning the possible mechanism of this biological control, Davey and Papavizas have produced data that largely confirm the explanation originally given by Blair that the activity of *Rhizoctonia solani* is temporarily checked

(a) by an increase in the carbon dioxide content of the soil and

(b) by a scarcity of available nitrogen in the soil solution. Both effects are produced by rapid decomposition of nitrogen, deficient organic matter in the soil. Davey and Papavizas showed that maximum inhibition of competitive saprophytic colonization of buck wheat stem segments by *Rhizoctonia solani* occurred in soils receiving amendments adjusted to C/N ratios 100-40. A lower degree of inhibition was provided by amendments with either higher (400-200) or lower (20. 10 and 5) C/N ratios. This result can be explained by postulating that nitrogen exerts two distinct and opposite effects upon the activity of R. *solani,* the first being predominant down to C/N ratio 40 and the second coming into action at C/N ratios of 20 and below.

The first or inhibitory effect can be attributed, to the action of nitrogen in enhancing the rate of decomposition of the organic material and hence in promoting a more rapid production of carbon dioxide. This increase in carbon dioxide output can be expected to closely fellow the trend in total microbial numbers from C/N ratios 400 down to 5. At C/N ratio 20, however, the second or stimulatory effect (with respect to *Rhizoctonia solani)* of nitrogen comes into force and the total decreasing effect of the organic amendment upon the fungus is suddenly decreased rapidly. This can be ascribed to the presence of the available nitrogen that can be directly absorbed by the mycelium of *Rhizoctonia solani,* thus removing the ,disability of nitrogen starvation suffered by the fungus in the presence of organic amendment adjusted to higher C/N ratios. This change is exactly where we might have predicted it to have occurred, that is, in the region of C/N ratio of 20. The C/N ratios of fallow soils and of microbial substance is generally around 20.

Mitchell reported that soil amendment with either chitin or laminarin at 1020 kg/hectare reduces incidence of bean root rot and also of radish wilt due to *Fusarium oxysporum* f. *conglutinans.*

Phytophthora root rot of avocado was controlled by Zentmyer by amending the soils with alfalfa meal-a material of low C/N ratio. Similarly, Mehrotra and Tiwari could successfully controlled the root *rot* of *Piper betle* by organic amendments and antagonists growing on organic amendments in the, soil amended with corn straw gave best control. Corn straw supplemented with nitrogen (NH_4NO_3) gave still better control. The effect of different antagonists growing on corn straw and til *(Sesamum indicum)* oil cake amendments showed that *Trichoderma viride* gave best control. Mehrotra and Claudius tried to eradicate the root rot and wilt diseases of *Lens culinaris* by organic amendments of the soil and also introducing antagonists in the soil along with the organic amendments. The wilt of lentil caused by *Fusarium oxysporum* f. *lentis* was checked with change of wheat straw, wheat straw plus ammonium nitrate, corn straw and sugar cane straw. The successful antagonists were *T. viride* and *Streptomyces gougeroti.*

Biological control in potato diseases of underground parts has been studied by Shrivastava and Saksena at Sagar. In soil amendments, the *Rhizoctonia* disease was controlled by nitrogen-rich amendments, such as the tops of bean plants. Carbon rich amendments, such as cellulose and sugarcane straw did not control the disease at all. This can be explained on the basis of increased activity of *Rhizoctonia solani* which readily breaks down cellulose. The wilt disease due to increased *Fusarium solani* could not be controlled in these experiments by organic amendments.

One good examples of biological control with green manuring is that of potato scab with a soybean cover crop in southern California. *B. subtilis* grew equally well in pots of soil amended with either green soybean or green barley tissue, although the bacterium multiplies in much greater amounts on soybean than on barley.

There are many examples where attempts have been made to infest the seeds or propagules with organisms with a view to reduce root disease. Some workers have however emphasized the non content effects of seed inoculation with respect to biological control. Chang and Kommedahl were successful in biologically controlling seedling blight of corn by coating seeds with *Bacillus .subtilis or Chaetomium globosum.*

i. *Hyperparasitism*

Certain fungi and bacteria act as parasites on plant pathogens. The genus *Cinccinobolus* includes species of fungi which attack members of the family, Erysiphaceae. *Cinccinobous* seldom reduces the attacks of the powdery mildews in nature because it very rarely occurs in large quantities. *Cinccinobous* can be grown on artificial media, however, and some good results have been obtained by artificial inoculation. Several workers have tried to utilize *Eudarluca australis* for controlling rusts. The fungus does control the disease under certain natural climatic conditions over which man has no control. As discussed by Chester extra inoculum of the hyperparasites will not be of much use as the weather, unless very favourable *to* the development of this hyperparasite, is unlikely to permit any marked control' of the disease. There are several other fungal species which attack rusts, such as *Tubercularia* spp. which attacks aecia and spermogonia of several rusts and *Verricillium hemieiae* on *Hemileia vastatrix.*

Many fungi are known to parasitize phytopathogenic nematodes. In 1934 Rozsypal and Schmidt reported a *Protomycopsis chytridiale* which attack *Heterodera schachtii.* In 1933, H. Goffart recognized that the eggs and embryoes of *Heterodera avenae* could be parasitized by an ascomycete, *Cylindrocarpon radidcola,* which is known to attack the underground parts of various cultivated plants. Hyphomycetes as nematode predators are numerous, and are frequently met with in nature.

Several attempts have been made to use fungi for the control of nematodes, such as Meloidogynes which attack pineapple in Hawai for nematodes parasitic on animals in France and for controlling the golden nematode of the potato in England. Some good results were obtained but practical use of these fungi in agriculture is still a far off possibility.

j. *Bacteriophages*

In 1917 d' Herelle presented proof of a transmissible lytic principle that acted on the I shiga bacillus, showing that bacteria can be diseased too. A number *of* phages have now been discovered for many phytopathogenic bacteria such as *Agrobacterium tume faciens, Erwinia aroidea, E. carotorora, corona. fadens, p, glycinea, P. tahaci. P. xanthoclona, Xanthomonas citri, X. phaseoli, X. solanacearum. X. oryzae,* etc. but not much success has been obtained *to* control these diseases at big scale with the help of their respective phages.

k. *Entomogenous Fungi*

There are many fungi which are capable of growing in the bodies of insects and thus destroying them. Scientists are looking for ways of protecting crops against insect pests. With the introduction of DDT, it was felt that a way had been found to destroy insects affecting crops. DDT and scores of other insecticides that, were introduced killed not only all the insects but also the predators, including fish in water (the insect enemies), with the result that in some cases more damage was done more than any benefit has occurred. Plant pathologists can use micro-organisms which are immune or at least not affected by pesticides for biologically controlling insects. One *of* the important fungus which has been utilized is *Entomophthora.* About 100 species are known, mostly as insect parasites. While some appear to be rather specific in their host range, others are capable of-attacking a wide range of hosts. Some species have been grown in culture.

PHYSICAL AND MECHANICAL CONTROL

The use of physical barriers such as row covers or trenches prevents insects from reaching the crop. Row covers can help prevent early-season damage to cucurbits by cucumber beetles, and plastic-lined trenches are effective in trapping large numbers of dispersing Colorado potato beetles in the spring and fall.

Cold storage is also considered a physical control and, although it does not necessarily kill the insect pests, it at least stops their development and further feeding on the stored crop. Other methods include hand picking of pests, sticky boards or tapes for control of flying insects in greenhouses and various trapping techniques.

CONTROL OF NEMATODES

Nematodes are very small roundworms. They live everywhere—in soil, plants, water, and animals, including humans. Nematodes feed on bacteria, fungi, algae, plants, and animals. A few thousand species attack plants; most nematodes feed on dead or decayingorgan-ic matter. Many nematode species are beneficial because they feed on bacteria, insects, fungi, and other soil pests. The life cycle of a nematode is egg, four juvenile stages, and adult. Adult females lay eggs that hatch into young nematodes, juveniles. Juveniles look like small adults. First and second stage juveniles are usually found still inside the egg case. For most plant-parasitic nema-todes, egg hatching takes place during the second juve-nile stage. After a total of four molts (shedding of their skin), the juveniles become adults. If males and females are present, they mate and produce more eggs. Typically, the life cycle of a plant-parasitic nematode is completed within 20 to 60 days.

FEATURES OF NEMATODES

Aquatic, triploblastic, unsegmented, bilaterally symmetrical, roundworms, that are transparent, colorless, usually bisexual, and worm-shaped (vermiform), although some can become swollen (pyroform). Only an estimated 3% of all nematode species have been studied.

HOME OF NEMATODS

Nematodes exist almost everywhere, including soil, plant and animal surfaces and interiors, decomposing life forms, and fresh and salt water and their subtending sediments. Nematodes can be grouped according to what they eat. The different groups are fungal feeders, bacterial feeders, predators, animal parasites, algal feeders,omnivores, and plant parasites.

MODE OF INFECTION

Plant-parasitic nematodes pierce plant cell walls with their stylet, which is formed by some of the mouth and esophagus parts. Then they pump up the plant cell just into their digestive system. These nematodes only feed on a certain part of the plant.

Most feed on the roots. Some of kinds of nematodes feed only on the outer tissue of the root, others penetrate more deeply, and some completely enter the host. There are different ways they harm the plants. Some induce their hosts to produce nutrients which the nematode can survive on, or enlarged structures in which the nematodes live, or both. Some produce metabolites which kill host tissue.

These activities all result in energy being removed from plants to support nematodes. Some plant-parasitic nematodes can cause severe damage that can result in unmarketablity, yield decrease, or even total crop failure.

SOME PLANT-PARASITIC NEMATODES ARE HARMFUL TO OREGON AGRICULTURE

Root-lesion nematodes (*Pratylenchus*) rip a hole in the sides of root cells and crawl inside. They move through the root, piercing, sucking, and leaving behind a trail of both cell-killing metabolites and eggs. Cell death results in brown lesions on the roots. Lesions begin on one side, but may encircle a root and thereby girdle it. The overall effect is a weak, shallow root system with a lot of dead areas.

Root-knot nematodes (*Meloidogyne*) enter roots as juveniles, select a feeding site of three to eight cells, and swell up in their chosen spot as they progress towards adulthood. They introduce hormone-like substances into the plant cells, causing the plant to swell in the area producing galls, or root knots. Males regain their slender profiles and leave the root at adulthood, but the fattened adult females remain inside.

They exude eggs into the soil, and after the juveniles hatch, the cycle begins again. Roots infested with root-knot nematodes usually have visible galls and may exhibit excessive branching. Parasitized plants may be weak and stunted. Root systems may be deformed, and underground organs such as potato tubers and carrot taproots may be damaged and unmarketable.

Pin nematodes (*Paratylenchus*) pierce root cells from the soil outside of the plant. Low numbers may appear inconsequential, but in high enough numbers, pin nematodes can damage crops. Ring nematodes (*Criconemella*) behave similarly. Little is known about ring nematode damage on many crop plants. Stubby-root nematodes (*Trichodorus* and *Paratrichodorus*) are also external soil-dwelling parasites. Besides piercing root cells and sucking out the contents, they can vector viruses such as the Tobacco Rattle Virus, which causes corky ringspot disease in potato tubers. Other virus-vectoring plant-parasitic nematodes are *Xiphinema*, daggar nematode, which can vector Tomato Ringspot Virus, a disease of many fruit crops, and *Longidorus* ("mint nematode").

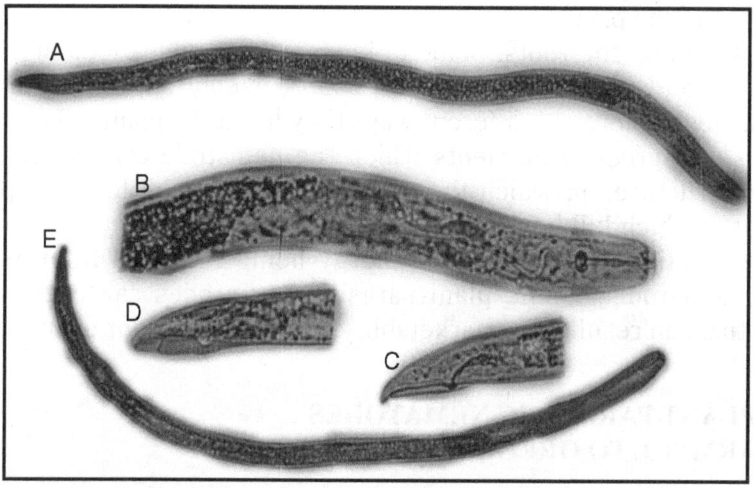

Fig. Vermiform (a) Adult Female Lesion Nematode, (b) anterior Region Showing Lobed Stylet, (c) Posterior of Female, (d) Posterior of Male Showing Spicule and Caudal Alae, (e) adult male lesion nematode

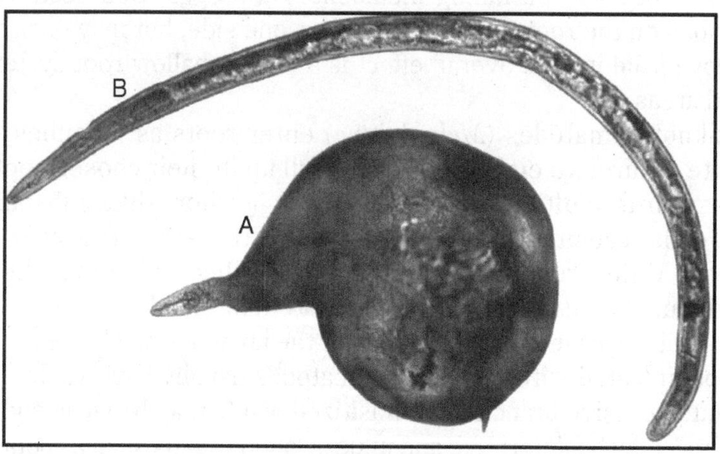

Fig. (A) Pyroform Female Root-knot Nematode, (B) Vermiform Male Root-knot Nematode

Nematode Organ Systems

Skeletal - exoskeleton (triple layered cuticle)
Secretory/excretory -

Nervous - nerve ring and longitudinal muscles
Digestive - stoma, esophagus, intestine,
Reproductive -
NO circulatory or respiratory systems - exchange of gasses through cuticle.

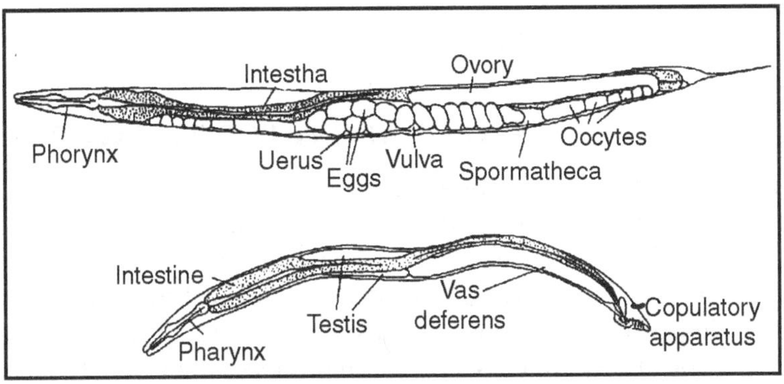

Fig. (Top) Hermaphrodite, (Bottom) Male

GENERAL PLANT PARASITIC NEMATODE LIFE CYCLE

5 Stages-4 Molts
Egg - J1 - M1 - J2 - (Hatch) - M2 - J3- M3 - J4 - M4 - Adult
In all plant parasitic nematodes the first molt occurs in the egg.

The emerging second-stage juvenile (J2) is the most common infective stage.

Length of life cycle of plant parasitic nematodes can be anywhere from 20-40 days (on average 25 days at 22°C).

NEMATODE MODES OF REPRODUCTION

Amphimixis - cross fertilization (males and females)
Parthenogenesis- nonsexual, males not required for reproduction(clonal)
Hermaphrodism - male and female organs in one individual(some change sexes)

Reproductive Potential (Egg production)

Anquina 25,000 eggs (2,000/wk)
Meloidogyne 500-1500 eggs
C. elegans 250-350 eggs
Ascaris 2 million eggs/day

NEMATODE GROUPS

1. Marine, Freshwater, Soil
2. Animal Parasites
A. Vertebrate

B. Invertebrate
3. Plant Parasites
Marine, Freshwater, Soil (Trophic Groups)
Microbivores ("free-living")
Fungivores
Omnivores (feed on multiple sources)
Predators (on nematodes, protozoa,etc.)

BIOLOGICAL CONTROL AGENT FOR NEMATODES

Too often, biological control agents have failed because they have been used before a basic knowledge of their ecology and biology has been established. The importance of such knowledge can be seen from work at Rothamsted aimed at the development of *V. chlamydosporium* as a biological control agent for root-knot nematodes. Stirling describes the design and methods used in the conduct of biological control experiments, which are more complex than those required to assess the efficacy of a nematicide. He identified five key aspects in setting up an experiment to evaluate a biological control agent and these are presented below in a slightly modified form.

- The test organism and any organic amendment should be applied at practical application rates; 0.1 percent w/w soil is equivalent to 2.5 tonnes/ha and should represent a maximum dose. Tests should always be performed in a non-sterilized soil with a natural residual soil microflora.
- Appropriate treatments as well as an untreated control should be included if the organism is added with a substrate. These treatments should include the substrate alone, the organism alone, and the autoclaved colonized substrate. Too often, untreated controls are compared only with large applications of the organism and substrate and this does not allow separation of the effects of the agent from the effects of the substrate. In several tests reported in the literature, application of the substrate alone has decreased nematode populations to the same extent as the substrate colonized by the agent, and there is no clear evidence of biological control.
- Population densities of the agent under test should be monitored to ensure that it has survived in soil throughout the period that activity against the nematode target is required. Such monitoring may require the development of selective media, which can be a difficult and time-consuming task.
- Nematode mortality caused by the organism under test should be measured to assess whether differences between nematode population densities in treated and untreated soil relate to the levels of kill caused by the agent. Infection levels are relatively straightforward to estimate

for most parasites, but repeated sampling is required to determine total kills. The effects of agents which produce toxins or have indirect effects on nematodes through competition, the modification of root exudates or the colonization of feeding cells, can only be measured by assessing their impact on nematode development.

- The impact of the soil environment, host plant and nematode should be tested as these are likely to affect the efficacy of the biological control agent, and could account for the lack of activity of potential agents in specific test conditions.

In a review of the literature less than 15 percent of experiments purporting to demonstrate biological control caused by *Paecilomyces lilacinus* satisfied the above criteria. Although this is an unsatisfactory situation that must be remedied, the difficulties in conducting carefully controlled and monitored experiments should not be underestimated.

Isolates of *V. chlamydosporium* differ markedly in their growth and sporulation *in vitro*, and in their virulence, saprophytic competitiveness and rhizosphere competence. Such differences between isolates of the same species of micro-organism are common and there is a need for simple laboratory-based screening methods to select the most promising isolates for further testing. *Verticillium chlamydosporium* isolates were collected from infested nematode females and eggs in suppressive soils around the world. Hence, they were isolated from the niche in which they would be required to be active as a biological control agent and from soils most likely to yield active isolates. Once isolated in pure culture using standard techniques, the *in vitro* growth requirements were determined.

Virulent isolates were selected by counting the number of nematode eggs parasitized after exposure to the fungus on agar in a standard test. Tests for proliferation in the rhizosphere and growth in soil were also used to select promising isolates. Although simple laboratory-based screens help eliminate many isolates that show insufficient activity to justify further testing, selected isolates will not necessarily be active in the field. As it may take 10 to 16 weeks to investigate adequately the performance of different isolates against cyst and root-knot nematodes in pot tests, relatively few can be screened. The method of mass culturing of *V. chlamydosporium* for experiments can have a marked effect on the subsequent survival and proliferation of the fungus in soil. Inoculum produced in shaken liquid cultures consists mostly of hyphae and conidia, which require an energy source to ensure proliferation in soil, whereas on solid media large numbers of chlamydospores are produced and these can be added to soil in aqueous suspension and rapidly establish the fungus.

The development of a semi-selective medium has enabled detailed studies to be made on changes in relative abundance of the fungus in soil and on roots. Some isolates of *V. chlamydosporium* may be extremely abundant in soil but unless they are capable of colonizing the rhizosphere they do not parasitize the eggs of root-knot nematodes.

Growth in the rhizosphere differs markedly between plant species, *e.g.*, tomato, cabbage and maize roots support much growth, whereas sorghum, pepper and cotton are poor hosts. Colonization by the fungus is confined to the rhizosphere and rhizoplane and there is no spread into root tissue; no lesions have been observed on roots grown in soil treated with *V. chlamydosporium* and there have been no detrimental effects on the growth of a range of crop species.

A tenfold reduction (from 10^4 to 10^3 chlamydospores/g soil) in the amount of fungus applied to soil had no effect on the extent of colonization in the rhizosphere; the ability to proliferate on the root surface where the fungus is required to control nematodes is an important characteristic which may allow significant reductions in the amount of inoculum applied to soil.

The efficacy of *V. chlamydosporium* as a biological control agent for root-knot nematodes is affected by three key factors: the amount of fungus in the rhizosphere; the rate of development of eggs in the egg masses; and the size of the galls in which the female nematodes develop. In large galls female root-kn2ot nematodes may produce egg masses which remain within the gall and are not exposed to parasitism by *V. chlamydosporium,* which is confined to the rhizosphere. Hence, *V. chlamydosporium* is less effective in controlling root-knot nematodes in heavily infested soils and on highly susceptible crops because large galls are formed on the roots and many eggs escape parasitism.

Verticillium chlamydosporium is unlikely to be useful in these situations where a grower would normally apply a nematicide. Also, at temperatures above 25°C eggs may complete their embryonic development and hatch before the fungus has completely colonized the egg mass; at 30°C about 30 percent of eggs of three root-knot species hatched and the second-stage juveniles escaped from the egg mass before the eggs were killed. These studies in pot tests, if supported by field experiments, help to define the conditions in which *V. chlamydosporium* might be used successfully for control of root-knot nematodes. It is only from such detailed studies that the limitations and requirements of the fungus can be assessed.

The Extent of Colonization of the Rhizosphere by *Verticillium chlamydosporium* on Several Plants Grown on Soil Treated with 5 000 Chlamydospores/g Soil and the Control of *Meloidogyne incognita*

Verticillium chlamydosporium is not a replacement for nematicides but, despite its limitations, it may be a useful management tool. In most tests biological control agents have been applied to protect susceptible crops. However, application of the fungus to a relatively poor host for the nematode on which small galls are produced so that most egg masses are exposed in the rhizosphere might provide more effective control; soil population densities of the nematode would be reduced to non-damaging levels before a susceptible crop was planted.

Host plant	Rhizosphere colonization (cfu/cm² root)[1]	Control (%)
Tomato	229	58
Maize	216	71
Cabbage	199	68
Potato	88	79
Tobacco	43	40
Pepper	29	36
Sorghum	25	0
Wheat	15	5

Note:

[1] cfu are the colony-forming units which develop on the semi-selective medium.

Bridge recommends the following rotation for the management of root-knot nematodes: susceptible host - poor host - poor host - resistant or non-host -susceptible crop. Application of the fungus before the first or second poor host may permit the shortening of rotations, or the replacement of the resistant or non-host crop, without increasing nematode damage to the susceptible crops. Currently, at Rothamsted Experimental Station, United Kingdom, poor hosts for root-knot nematodes are being screened for their ability to support *V. chlamydosporium* in their rhizospheres. It is hoped that by combining the host plant and the biological control agent, more effective and consistent control can be achieved.

The Next 20 Years

The past 20 years have seen a significant increase in the number of scientists involved in research on the biological control of nematodes. Surveys and empirical tests are being replaced by quantitative experimentation and basic research on the modes of action, host specificity and epidemiology of selected organisms.

Such basic information is essential for a realistic appraisal of the impact of molecular biology on the improvement of microbial agents and monitoring the spread and survival of released organisms, and for the development of rational strategies for control.

Current experience suggests that biological control agents will not replace the use of nematicides but, integrated with other control measures including chemicals, they could play an important role in the development of integrated control strategies in both developed and developing agriculture. The urgent need to reduce the dependence on nematicides should provide the necessary impetus for the considerable amount of research and development still required to ensure the successful use of such agents.

Pathology

Bailey and Gordon reported that *R. culicivorax* depleted host metabolites, reduced fat body and other host storage tissues while accumulating stored materials in their trophosomes, and thus inhibited the development of imaginal disks in the host mosquito. Hemolymph proteins were depleted sixfold in mosquitoes.

Mermithid parasites of larval diptera probably resemble *M. nigrescens* in obtaining dietary amino acids by stimulating the catabolism of proteins within the host fat body. In addition to the reduction of most protein fractions, significant decreases also occurred in hemolymph glucose in parasitized blackflies; however, blood trehalose concentrations were not affected. Glycogen reserves were shown to be similarly reduced in the fat body of mermithid-parasitized simuliids. Gordon *ET al.*found that parasitism did not significantly affect either the overall concentration of lipids or relative proportions of the lipid fraction in the hemolymph of mosquitoes. Mermithid parasitism of mosquitoes and simuliids caused almost complete degeneration of host fat body tissue. All storage metabolites, including glycogen, within the fat body may directly or indirectly be utilized by the developing nematode.

Responses are somewhat different in larger hosts with longer periods of parasitism. Protein in the fat body of grasshoppers parasitized by M. *nigrescens* was depleted after 2 weeks; corresponding reductions in proteins were not recorded from the hemolymph until I week later. The third week of infection represented a period of maintenance and reconstruction of fat body soluble proteins, but corresponding effects were not found for hemolymph proteins until the fourth week of infection.

Gordon and Webster found that the overall level of amino acids in the hemolymph of adult female grasshoppers was not affected by mermithid parasitism, but the content of amino acids within the fat body was significantly reduced. Parasitism by *M. nigrescens* resulted in glucose levels remaining low for the first 4 weeks of parasitism relative to trehalose, but no significant difference for glucose levels when compared to unparasitized controls. Trehalose levels dropped and remained low for much of the parasitic period. These data suggested that the nematode modifies its host's metabolism to favour production of small molecules such as glucose and amino acids from carbohydrate and protein reserves and that the host cell's ability to sequester nutrients is impaired during parasitism.

Craig and Webster found that ecdysone levels in locusts were unaffected by *M. nigrescens* and attributed the inhibition of molting in parasitized insects to depletion by the nematode of precursors required by the host for protein and cuticle synthesis. However, ovaries of parasitized locusts were unable to sequester vitellogenic proteins available within the hemolymph, suggesting endocrine dysfunction in the host.

Major modifications of the host's metabolism by mermithid parasitism are manifested in host tissue degeneration and retarded development including resorption or suppression of oocyte development. As a result, hosts are usually prevented from maturing to the adult stage or when they do they are generally unable to-, reproduce.

Hosts are occasionally able to prevent parasite development by either cellular or humoral responses. The most common cellular response is encapsulation. Poinar *et al.* reported that *Culex territans re*sponded to the development of parasitic juveniles of *R. culicivorax by* covering the 2- to 3-day-old parasites with blood cells. These nematodes failed to complete development. A more frequent condition, melanotic encapsulation, was reported in *Diabrotica* beetle larvae attacked by F. *leipsandra.* Host hemocytes lysed on the cuticle of the nematode soon after it entered the host, and within 6-8 hours an inner layer of melanin had formed around the nematode.

Similar responses were observed for larvae of *Psorophoraferox, Aedes triseriatus, and Anopheles quadrimaculatus* to the *R. culicivorax. How*ever, some 70 species of mosquitoes have not elicited this type of defence response to *R. culicivorax*. Recently, Gaugler *et al.* reported an encapsulation response when *Empidomermis* sp. entered larvae of *Aedes stimulans* if the nematodes were unable to migrate to the host's head and enter central nervous tissue.

After pupation, the nematodes dropped back into the hemocoel where they no longer elicited a host defence reaction. This response may be common in mermithids that enter the larval stage but fail to mature until the host reaches the adult stage. Also, Harlos *et al.* noted that *Culicimermis* sp. entered the nerve tissue of *Aedes vexans* larvae and later developed in adult hosts; Petersen (unpublished data) observed a similar response in *Ae. sollicitans* when parasitized by *P. culicis.*

Humoral responses, non-cellular components of the insect's hemolymph which have an adverse effect on nematode development, may account for the host resistance observed in *Aedes intrudens* and *Aedes provocans* to an *Empidomermis* sp. when no discernible host response was observed, which was unlike the encapsulation response this nematode elicited in *Ae. stimulans*. Another type of humoral response may be that observed when blackflies were subjected to the mosquito mermithid *R. culicivorax.* The mermithid initiated development in early-instar; Iackflies, and direct host reaction to this unnatural parasite was not observed; eventually, however, both nematode and host died.

Physical and behavioural responses also influence parasitism by mermithids. Host age, especially in aquatic insect hosts, has been shown to have a considerable effect on host susceptibility. Refractiveness because of age appears to be caused by the thicker cuticle in older larvae. In the laboratory, as a result of mass-rearing practices for *R. culicivorax,* where exposed but uninfected hosts were retained to supply the next generation, the host mosquitoes were found to have become measurably less susceptible to the nematode after some 100 generations. No evidence of humoral defence mechanisms was found and the mode of defence was not determined. In a similar buildup of refractiveness in *An. quadrimaculatus to S. Petersen,* it was reported that less susceptible hosts were more active and aggressive in attempting to remove attacking preparasites.

Natural Population Regulation

Mermithids generally have localized, discontiguous distribution. This is especially true for mermithids that develop in larval hosts. *Romanomermis culicivorax* was found in only seven pools out of several thousand examined in southwestern Louisiana. This is understandable because the hosts are killed prior to emergence from the habitat, thus greatly reducing the nematodes ability to disperse.

Generally, observations of mermithid parasitism are based on isolated collections and often give the impression of higher parasite activity than actually exists. In extended studies, levels of parasitism are usually found to vary greatly. In a 27-month survey of five habitats containing populations of *R. culicivorax,* mean parasitism of *Anopheles crucians* for individual habitats ranged from 8 to 42 per cent, with parasitism of individual populations as high as 80 per cent. Similar but lower levels of parasitism were observed for several other mosquito species.

Similarly, Phelps and De Foliart observed parasitism of blackflies, *Simulium vittatum,* in four Wisconsin streams over a 1-year period. Parasitism typically ranged from 10 to 90 per cent and the authors suggested that localized parasite populations probably attained levels of parasitism sufficient to virtually eliminate some host populations. Similar population reductions of blackflies were reported by Welch and Rubtsov.

Parasitism of adult aquatic insects usually occurs over a wider geographical area and often reaches very high levels but fluctuates greatly over time. Steiner observed 80 per cent parasitism of adult *Ae. vexans in* British Columbia in 1920, but only 20 per cent the following year. Trpis *et al.* reported 100 per cent parasitism of the same host in the same locality in 1967. Similar observations were made on parasitism of *Ae. sollicitans* adults by *P. culicis* in southwestern Lousiana. During an entire year only collections made in January failed to produce infected hosts; mean monthly parasitism ranged from 0 to 48 per cent with individual populations reaching 96 per cent. During a period of high parasitism, a survey was made along 80 km of coastline and revealed parasitism ranging from 48 to 94 per cent for seven populations.

Similar isolated high levels of parasitism are frequently observed in terrestrial insects. Christie reported a heavy outbreak of grasshoppers in Wisconsin in 1923-1925 with a correspondingly heavy infestation *of M. nigrescens.* The author concluded that the parasites were an important factor in terminating the grasshopper problem. Similarly, Mongkolkiti and Hosford reported that a population of the grass*hopper Hesperotettix viridis pratensis* was totally destroyed prior to egg lying by *M. nigrescens.* Other species of grasshoppers in this population were only partially affected or not affected by the mermithid.

Though reports of natural insect control are impressive and suggest the potential of mermithids as biological control agents, these parasites generally do not influence significant control over host populations. In a review of blackfly-mermithid literature, Molloy reported that parasitism in most blackfly populations is moderate, ranging from 3 to 15 per cent, and is perennial with only rare and highly localized epizootics. This pattern probably holds true for most mermithid-host associations.

Safety

As previously mentioned mermithids are generally specific to one or a few species of insects and are rarely found in other insects. As a result, little concern has been shown for potential danger of these parasites to mammals. To date, only limited mammalian safety studies have been made. Ignoffo *et al* reported that when suckling and adult mice and adult rats were subjected to either *per os,* intranasal, intraperitoneal, or dermal challenge of *R. culicivorax,* their body weight gain and histologies were identical to those of untreated animals. Immunodepressed rats also were not susceptible. Similar findings were reported for *R. iyengari in* India. Similarly, studies in the People's Republic of China showed that no pathology was apparent in dermally challenged suckling mice, in orally exposed adult rats, or in three species of fish when exposed to *R. jingdeensis*.

Mermithids have occasionally been reported as accidental parasites of man. Poinar reviewed all such cases in the literature and concluded that in most cases the data were insufficient to ascertain if human parasitism actually occurred. He further concluded that the question of accidental human infection by mermithids cannot be definitely answered at this time and should only be accepted as fact when proven experimentally or when parasites are found developing *in situ* in the human organism. Further, the Environmental Protection Agency has exempted mermithids from safety regulations.

Mass Propagation

Mass production systems at economical costs are essential if a given biological control agent is to be effectively used. Mermithids take in their nutritional, needs directly through the cuticle, making them especially difficult to propagate by *in vitro* methods. Therefore, pest control attempts have been limited to those mermithid-host systems that are conducive to *in vivo* rearing.

In Vivo **Propagation**

The *in vivo* rearing of mermithids of terrestrial insects is especially difficult because of the protracted life cycles of many of the hosts. Though numerous mermithid species have been encountered, laboratory culture has been limited to *M. nigrescens and F. leipsandra.* Laboratory infections are relatively easy with *M. nigrescens.* Gordon and Webster collected adult females of *M. nigrescens* from the field, allowed the females to oviposit, concentrated the eggs on moist filter paper, and stored the eggs at 5°C until needed. To infect hosts, eggs were transferred onto small pieces of grass previously coated with an adhesive material. The grass was readily eaten by host grasshoppers. This method permitted extensive research on the physiological effects of mermithid parasitism on a host system. The long developmental period of the nematodes during both the parasitic and free- living stages (I year) has prevented meaningful use of this nematode in field trials.

Creighton and Fassuliotis were the first to give a detailed report of the laboratory culture of a mermithid of terrestrial insects. Preparasitic juveniles of *F. leipsandra* and the larvae of the cucumber beetle, *Diabrotica blateata,* were placed in holes in soil in clay pots. The pots were sealed and held for parasite emergence. Ratios of 1000 nematodes per 100 host larvae were sufficient for maximum recovery of postparasitic nematodes. The method has the potential of producing between 80,000 and 240,000 nematode eggs per experimental unit. From this basic procedure, Creighton and Fassuliotis developed an *in vivo* mass rearing system for *F. leipsandra.* As eggs were laid, they were concentrated, sterilized by immersion for 10 minutes in 0.28 per cent sodium hypochlorite, and rinsed.

The eggs were transferred to a Baermann funnel and allowed to mature and hatch. First instar larvae of the banded cucumber beetle were exposed to the preparasites in clay pots as previously described. Twenty-four pots were placed in a plastic box and held 4-5 days at 24-27°C. The contents of the clay pots were then transferred to porous baskets. The baskets were placed at one end of a larger container with sand at the other end. As the host larvae matured, they migrated to the sand and pupated. After about 10 days the nematodes began to emerge from their hosts. The postparasites were then transferred to a container of aerated water until the nematodes molted to the adult stage and began to deposit eggs (7-10 days). Each larger container (37 x 27 x 10 cm) produced an average yield of 1.9 X 106 nematode eggs. The system can produce about 5 x 106 eggs week- I at a cost of $0.40 (US$) per I 000 eggs including supplies and labour for producing the host insects.

Although mermithids are relatively common in aquatic insects, especially in the Simuliidae, Chironomidae, Ceratopogonidae, and Culicidae, many of these host species cannot be successfully maintained in laboratory colonies or cannot be maintained in large numbers.

Also many mermithid species undergo egg diapause or have an asynchronous hatch which prevents their mass production. As a result, few aquatic mermithid species have been studied extensively. The midge parasite, *Hydromermis conopophaga,* has been readily maintained in colony because the host, *Tanytarsus* midges, can be easily maintained. However, *H. conopophaga* has never been mass produced or released in the field. Similarly, the mermithid *Limnomermis rosea,* a parasite of chironomid larvae, has been reared through several generations but has not been released in the field.

Muspratt was the first person to develop, maintain, and report a laboratory culture of a mermithid nematode. He collected *0. muspratti (Agamomermis)* from naturally infected host mosquito larvae and placed the nematodes in containers filled with moist sandy soil. The soil was allowed to dry out over a period of several months. After 11-12 months some of the soil was placed in a container of water and first instar mosquitoes added. Seventy to 80 per cent of the host larvae contained nematodes. The original work by Muspratt though very basic provided the direction for future work with mosquito mermithids. Subsequent studies with 0. *muspratti* have refined the rearing technique and defined biological limitations. A mass production system for *0. muspratti has* failed to develop, even though a suitable laboratory host system is available, because the eggs of this species failed to hatch at the same time although they appeared to mature. Because of the non-synchronous hatch, cultures were observed to produce infective stage nematodes after periodic floodings for over 5 years (Petersen, 1981).

The mermithid, *S. peterseni,* a parasite of *Anopheles* mosquitoes, has been maintained in colony for over 10 years. Though this nematode appears to be an infective parasite, and is easily maintained in laboratory colony, economical mass production systems have been difficult to achieve because the only hosts are *Anopheles mos*quitoes, which are less suitable for mass rearing systems than are certain *Aedes and Culex* species.

The most efficient and effective *in vivo* rearing system for mermithids is that developed for *R. culicivorax.* The primary techniques were developed in 1971. This initial system called for the exposure of 20,000 first-instar *Cx. quinquefasciatus* larvae in 136 x 52 x *5* cm galvanized 'Lrays to preparasites at a parasite-host ratio of 12: 1. The hosts were then fed a regimented diet for 7 days. After this period the culture trays were drained and the mosquitoes concentrated. Pupae of uninfected mosquitoes were separated using a chilled water procedure, and the infected mosquitoes were placed in 36 x 25 x 10 cm trays and held for nematode emergence.

Postparasitic nematodes were then washed and 10-15 g placed in paraffin-coated aluminum pans (22 x 33 x 5 cm) which contained clean, coarse, sterile sand and water. After 3 weeks the free water was removed and the cultures stored an additional 4-15 weeks before use. When preparasitic nematodes were needed the cultures were flooded with chlorine-free water to stimulate nematode hatch. Nematode cultures averaged 1.91 million preparasites (45-60 per cent of total yield) if they were flooded for the first time when they were 11- 19 weeks old. Total yields were highest (5.32 million preparasites) when cultures were flooded for the first time when 8-10 weeks old and then at 3-4 week intervals thereafter.

It has since been determined that yields can be increased by having a density of 24 postparasites CM-2 in the culture trays. Yields of preparasites were tripled by simply setting up three cultures (22 x 33 x 5 cm), each containing 5 g of nematodes, instead of the usual method of one culture containing 15 g of postparasites. The longevity of cultures could be extended for up to 9 weeks if the cultures were allowed to mature at ambient temperature and then held at 5-IOC. Also, immature cultures survived best at 15-20'C. Further, Sterling and Platzer reported that by adjusting the pH of water to 4.5 in R. *culicivorax* cultures, the nematophagous fungus *C. anguillulae* could be effectively controlled.

Chapman and Finney reported that mature eggs of *R. culicivorax* can be stored for 12 weeks at IO°C and 21 weeks if stored during early development at 15°C and later transferred to IO°C. They also reported that eggs of *R. culicivorax* placed in a damp fluoroform product could be shipped with only a 10 per cent loss. The technique has the advantages over the standard shipment of sand cultures of being about eight times lighter, and costing less to ship in addition to reducing shipping losses.

In an actual test of the mass propagation system, sufficient *R. culicivorax* were reared to treat 144,000 m2 of breeding area in El Salvador. The necessary inoculum required the exposure of 1.6 million first-instar *Cx. quinquefasciatus* to 137 million preparasites (I: 14 ratio) each week for 6 weeks. The system produced an average of 13.7 g (about 2000 g- 1) of postparasitic nematodes per rearing tray (20,000 mosquitoes), a total of 6392 g for the 6-week period and 435 cultures.

Bibliography

Mestry Urmila D., Mohite Vaishali, Mujawar Shabnam M :*Grapes and Their Management : T.V. Sathe, Lad Jothi S.*, Kumbhar Pravin D., , Shruti Pub, 2010.

M.K. Sheikh, N. Manjula and S.R. Mulla: *Recent Advances in Grapes* : , Daya, 2014.

S G Borkar: *Diseases of Grape : Their Forecasting and Control* , Pointer Pub, 2007.

K.G. Shanmugavelu: *Grape : Cultivation and Processing* , Agrobios, 2008.

T.V. Sathe, Lad Jothi S., Kumbhar Pravin D., Mestry Urmila D., Mohite Vaishali, Mujawar Shabnam M *Grapes and Their Management* , Shruti Pub, 2010.

S.J. Singh: *Virus Viroid and Phytoplasma Diseases of Grapevines* , Westville Publishing House, 2014.

Aruna A Birbal Jain: *A To Z of Health Challenges After Thirty : A Practical Guide To Your Health Needs*, Macmillan India, 2006.

Arvind Kumar Goel: *A College Text Book of Health and Hygiene*, ABD, 2005.

C S Jain: *A Complete Book on Health and Nutrition*, Cyber Tech, 2009.

C. Kaetsch : *Crop Science and Land Use for Food and Bioenergy*, Agrobios Publication, Delhi, 2010.

Deepa Joshi: *Microbes and Malignancy*, Oxford Book Company, Delhi, 2012.

Dilip Shah:*Co-Operativization Liberalization and Dairy Industry in India*, ABD Publishers, Delhi, 2000.

Dinesh Arora: *Biotech's Dictionary of Dairy Science*, Biotech Books, New Delhi, 2003.

Doreen Virtue : *Constant Craving : What Your Food Cravings Mean And How To Overcome Them*, Hay House India, Delhi, 2011.

Esmw Floyd: *1001 Little Health Miracles*, Carlton Publishing, 2009.

G. Sutherland Thomson:*Grading Dairy Produce*, Medi World Press, Kerala, 1995.

Gaurav Bhardwaj: *Applied Dairy Microbiology*, Sonali Publications, Delhi, 2012.

Gilles Dussap : *Current Topics on Bioprocesses in Food Industry*, Asiatech Publication, Delhi, 2006.

H.A. Modi:*Dairy Microbiology*, Aavishkar Publishers, Delhi, 2009.

Harmeet Singh: *Dairy Farming*, APH Publication, Delhi, 2011.

J David: *Contemporary Trends in Dairy Plant Management*, Kitab Mahal Agencies, Delhi, 2007.

K. Singh:*Dairy Microbiology*, Oxford Book Company, Delhi, 2012.

Kalpana Swaminathan: *A Compendium of Family Health*, Rupa, 2005.

Kate Barber Winton : *A Handbook of Structure and Composition of Foods*, Agrobios Publication, Delhi, 2013

Satish Sonkar: *World History of Physical and Health Education: Modern Period*, ABD Pub, 2005.

Sieger, Robin: *42 Days to Wealth, Health and Happiness*, Random House India, Jan-06.

Sita Ram Sharma: *A Handbook Of Health Education,* Sarup Book Publishers, 2009.

Sivakumar P.S. : *Agri Food Crops : Processing Value Addition Packaging and Storage*, New India Publishing Agency, Delhi, 2012.

Syed Aftab Iqbal and Neelofar Iqbal: *Advanced Food Chemistry*, Discovery Publishing House, Delhi, 2011.

Tharakan : *A Young Hotelier's Guide to Food and Beverage*, Tata McGraw-Hill Publication, Delhi, 2005.

U.K. Mishra, M. Singh and D.K. Sharma: *Organic Dairy Farming*, Satish Serial Publication, Delhi, 2011.

Urvashi Nandal : *A Handbook of Foods and Nutritional Biochemistry*, Agrobios Publication, Delhi, 2013.

Vanete Thomaz Soccol and Christian Larroche: *Advances in Bioprocesses in Food Industry*, Asiatech Publication, Delhi, 2011.

W. Merback and A. Veha : *Crop Science and Technology for Food Security Bioenergy and Sustainability*, L. Bona, J. Pauk, , Agrobios Publication, Delhi, 2012.

Kavita Marwaha : *Control and Analysis for Food and Agricultural Products*, Daya Publication, Delhi, 2010.

M K Rao : *Food and Dairy Microbiology*, Manglam Publication, Delhi, 2007.

M.M. Padhi: *A Hand Book of Common Siddha Kayakalpa (Rasayana) Herbs for Healthy Life*, Central Council for Res, 2008.

Mangala Kango : *Milk and Milk Products*, RBSA Publication, Jaipur, 2006.

Index